Palgrave Studies in Ethics and Public Policy

Series Editor
Thom Brooks
Durham Law School
Durham University
Durham
UK

Palgrave Studies in Ethics and Public Policy offers an interdisciplinary platform for the highest quality scholarly research exploring the relation between ethics and public policy across a wide range of issues including abortion, climate change, drugs, euthanasia, health care, immigration and terrorism. It will provide an arena to help map the future of both theoretical and practical thinking across a wide range of interdisciplinary areas in Ethics and Public Policy.

More information about this series at
http://www.springer.com/series/14631

Simon Reader

The Ethics of Choosing Children

palgrave
macmillan

Simon Reader
Keele University
Newcastle-under-Lyme, UK

Palgrave Studies in Ethics and Public Policy
ISBN 978-3-319-59863-5   ISBN 978-3-319-59864-2   (eBook)
DOI 10.1007/978-3-319-59864-2

Library of Congress Control Number: 2017943660

Cover illustration: © John Rawsterne/patternhead.com

Printed on acid-free paper

This Palgrave Macmillan imprint is published by Springer Nature
The registered company is Springer International Publishing AG
The registered company address is: Gewerbestrasse 11, 6330 Cham, Switzerland

For Alice Mary Reader

Acknowledgements

This book is based on research carried out at Lancaster University, for which I am grateful to acknowledge funding from the Arts and Humanities Research Council. Prior to this, postgraduate study funded by the Wellcome Trust set me on the road to this point, so I extend thanks to both funding bodies for making this possible. Thanks also to Brendan George at Palgrave Macmillan for investing in me, and to April James for her cheerful advice and correspondence.

Whilst at Lancaster I benefited from the guidance, expertise and reassurance of many colleagues, among whom I'd like to extend particular thanks to Alison Stone for her kindness, time and wisdom. I should also like to acknowledge the support of David Archard, along with the contributions of Mairi Levitt, Garrath Williams, Gavin Hyman, Imogen Tyler and Linda Woodhead at Lancaster University. I'm also grateful to the Gender and Women's Studies community who offered me a second home at Lancaster—and to new friends and colleagues for the warm welcome at Keele University.

Some incredible friends set me on this path and kept me steady when I needed it, and so I owe them all a debt of gratitude. In particular, Steve Crawford, who persuaded me to give academia another go, Andy and Maja Smith-Bugge, Rachel and Fred Binley, the Manchester crowd (who know who they are) and Captains Ali Hanbury, Rachael Eastham and Jez Mort for keeping me afloat. Thanks also to the Marketing Team at the University of Cumbria for being such good company and putting up with me for a year.

I am thankful to have a wonderful family, who have been an unwavering source of strength, support, love, joy and encouragement: my incredible, loving parents Christine and John Reader; Kate, John, Finn and Lucy; Tom, Steph and Maxwell. Likewise, huge thanks also to my godfather Andrew Buckley, mother-in-law Jennifer Boon, and to Ben, Erin and Izobel, I am so grateful for your support. I'm fortunate enough to have two formidable grandmothers in Mary Reader and in Florence Brierley, who provided a haven and another home in the north for so many years. I'd also like to name my late grandfather Eric Brierley, as an abiding influence, example and inspiration.

All of this is possible because of one man: my patient, talented, inspirational husband Russell Reader. His faith, passion, intelligence, love and gumption are the reasons any of this got done. All of the flaws in this book are mine; anything of worth or quality in the pages that follow I dedicate to him.

Contents

1

Bioethical Burdens of Proof

Abstract Modern and developing forms of selective reproduction have created new procreative choices for prospective parents which pit inclusive social values against personal parental preferences for one's own child. A liberal eugenic current of thought has become predominant in mainstream bioethics, commending the prenatal selection of one's children, where possible, over wider social or moral concerns about the nature of the world this creates. This opening chapter accounts for the development of a liberal eugenic orthodoxy, detailing the presumptions of liberty, moral considerability and the parity of genetic and environmental influences that support a permissive procreative liberty.

Keywords Bioethics · Reproduction · Liberty · Choice · Selection

Introduction

In one of the most remarkable books of recent years *Far from the Tree* (2012), psychologist and writer Andrew Solomon detailed his research with over three hundred families in which parents had been profoundly challenged by the exceptional conditions of their children. From autism

© The Author(s) 2017
S. Reader, *The Ethics of Choosing Children*, Palgrave Studies in Ethics and Public Policy, DOI 10.1007/978-3-319-59864-2_1

to transgenderism, deafness to criminality, Solomon investigated the relationships engendered by the unexpected, unpredictable and often unwanted alterity of our offspring. During his interviews, the conversation occasionally led to the question of a hypothetical 'cure' for whatever the difficulty may be, such as a couple whose son David was born with Down's syndrome. His mother says: "For David, I'd cure it in an instant; but for us, I wouldn't exchange these experiences for anything. They've made us who we are, and who we are is so much better than who we would have been otherwise" (217). Likewise his father declares: "For David, I'd do it. But the diversity of human beings makes the world a better place, and if everyone with Down's syndrome were cured, it would be a real loss. The personal wish and the social wish are in opposition" (217). Both parents wish, as we might expect, that if they could make it easier for their child to be in the world, they would do so. But reflecting on the construction of their own identities and the kind of world they want to pass on, they couldn't wish for the elimination of the condition itself.

In this book I want to address this question of the kind of world we wish to pass on, and its connection with the new technologies of reproduction that allow for the selection of those who will inhabit it. As such, it strays beyond the boundaries of reproductive bioethics, which ordinarily limits itself to the satisfaction of personal wishes, to consider the wider social significance of the new predicament of generation for generations. The aim is not to advance a vision of the world that thereby ought to be honoured in individual reproductive choices, but rather to create the space for acknowledging and expressing the pertinence of such visions to those choices. The opposition between the personal and the social wish described above conveys the anguish of reproductive decision-making where prospective parents entertain competing visions between aspirations for their child and aspirations for the world. As we know, the scope of such decisions now affords the prenatal selection of children for the possession or avoidance of certain conditions—although there is no 'cure' for Down's syndrome, it is now possible for prospective parents to avoid giving birth to a child with the condition. How do we make sense of a realisable personal preference to have a child without Down's syndrome, for example, when we also hold a sincere social wish that future generations still be hospitable to people with the condition? As one generation gives rise to the next, what changes with the shift from giving birth to strangers to choosing our children? And what could be morally and politically important about preserving ourselves and future generations from those choices?

The terms of the bioethical debate surrounding the use of Assisted Reproductive Technologies (ARTSs) have been staked out over the last few decades, with the emergence of a basic liberal presumption in principle favouring an individual freedom to choose, where possible, the desired or most desirable characteristics of one's child (see, for example, Buchanan et al. 2000; Agar 2004). Yet reservations or objections about this presumption can then be uncritically and unfortunately cast together as conservative or paternalistic attempts to curtail the freedom of others. It seems to me that this isn't a helpful way of preserving the genuine tension between the personal wish and the social wish in reproductive decision-making, or always a fair characterisation of misgivings about the ethics of choosing children. Declining such a choice can be depicted as a negative refusal to embrace the means to secure the most beneficent outcome for one's child or children; yet one could also think of it as expressing a radically liberal and positive ethical gesture of welcome and responsibility for whoever may come into existence. This ambivalence arises not only because of the personal interest we have in our own offspring, but owing to the fact that bringing individuals into existence raises particular kinds of moral questions that are unlike others, since only in these choices is the creation of brand new human beings at stake. Thus, it is held that one can consistently abhor discrimination in principle, but exercise it when bringing individuals into existence since there are no personal harms visited upon the resultant individuals concerned.

This permissible discrepancy between moral principle and reproductive practice feels problematic; where we can act with new techniques of reproduction so that moral values can be trumped by parental preferences, it is difficult to imagine that something is not also lost, and that a debate which doesn't register this isn't missing something. Of course, we can expect that most reproductive decisions be made with the resultant child's happiness and welfare at heart, just as so many parental choices around rearing existing children are ordered. We live in an age of unrivalled exaltation of our children; British Conservative MP Rory Stewart has lately declared "ours the first civilisation to find its deepest fulfilment in its descendants. Our opium is our children" (2013). But while within families we may become able to choose our own individual children, within generations we cannot opt out of the fact that we will deliver to the next a common location in history, together with the limits of their potential experiences and the shape of the world and values we leave to it. By considering the new technologies of selective reproduction together with what Mannheim called the Problem of Generations, I hope to articulate what may be missing from the debate, arguing that a principle of procreative beneficence alone does not capture the ethical significance of generation for generations, and continues to overlook the sexed phenomenology of birth in a manner that does violence to those who give it.

It is the curious, permissible but counter-intuitive discrepancy between moral principle and reproductive practice that suggests this can be a dangerous place to be subject to such practices, or certainly the narratives and norms that are instituted around them (Rapp 1999). With the advent of what Rothman has called the tentative pregnancy (1993), accounts of expectant mothers being pressured to undergo prenatal diagnostic testing or terminations are attested to in news reports (Miller 2016), and on websites such as Downs Side Up (2017) and Jacob (2017). In Sally Phillips' BBC documentary *A World without Down's Syndrome* (2016), one mother claimed that she had refused prenatal testing purely in order to keep the information from medical professionals, who she felt would harass her to terminate if the result was positive for Down's. One object of this book is improving the experiences of such women, so that the clinical journey through conception and pregnancy preserves genuine space for the principled refusal of such information without the fear of being impugned or coerced. On this theme, it is important to acknowledge my own position in relation to writing about pregnancy and childbirth, namely, as a childless man. Regretfully, as Kelly Oliver has observed, the vast majority of philosophical literature on genetic enhancement and reproductive choice is "authored primarily by prominent mainstream analytic philosophers (who—gender-testing aside—appear to be almost exclusively male)"

(2013, 221). Whilst it is my privilege to be male, the following pages should dispel any suspicion of my being mainstream, analytic—or prominent.

Chapter Summary

The remainder of this chapter will introduce the idea of selective reproduction in its present and prospective guises, which are distinct from earlier forms by being less harmful to the mother, child or foetus, and by offering more immediate and specific contrivance of the desired child against an informed choice about possible others. I will go on to explain how views on the use of these techniques of selective reproduction, in practice and in contemporary orthodox Anglo-American bioethics, have moved to a broadly liberal eugenic consensus. This is by way of levelling burdens of proof upon three key presumptions, against which those who object to the use of selection and enhancement technology in reproduction are answerable. Objections to the unfettered use of new forms of selective reproduction and enhancement must demonstrate how they visit harm upon morally considerable lives, or determine more perniciously the lives of future children any more than environmental influences exerted by parents and care-givers. Against this considerable burden, no demonstration of such harms or moral differences has proved convincing.

Chapter 2 introduces the discourse of the gift in relation to reproductive technologies, taking Michael Sandel's argument against enhancement as the most developed account of an objection to selective reproduction on the grounds of life's giftedness. This account, developed further by Michael Hauskeller, fails to convince against the charge of identifying a non-theistic giver of such a gift, or a compelling moral difference between genetic and environmental interventions that influence the make-up of future people. We turn then to consider whether, quite apart from life being given, it can be considered a good. Addressing first those extreme and unfortunate cases of so-called 'wrongful life' alongside a discussion of the non-identity problem, it seems clear that some lives cannot be considered good on any definition even though the alternative would have been non-existence. This leads to a discussion of David Benatar's anti-natalist argument that it is better never to be brought into existence; that life is not a good at all under any circumstances so could hardly be regarded as a gift. Finally, I consider Julian Savulescu's principle of Procreative Beneficence, which he pits directly against Sandel's argument and the intuition that life should be considered a gift. Savulescu argues for the parental obligation to use reproductive technologies to select the child with the best chance of leading the best life. I contend that although Savulescu dispenses with the language of gifts, he retains the notion in morally entreating parents to give life as beneficently as possible, just as Benatar's anti-natalism warns against procreation altogether because life is never a good thing to have been given. The narrative emerges that the virtues of gratitude and receiving gifts espoused by Sandel no longer apply now parents are able to exercise choice through assisted reproduction; rather the virtues become those of parental gift-giving, enabled by the prenatal diagnostic technologies that allow us to select beneficently for future individuals.

Chapter 3 opens with a discussion of the common rendering of natural reproduction as a 'creation lottery' in bioethical discourse, such that techniques of enhancement and selective reproduction are lauded as enabling the colonisation of the natural by the just. John Harris and Julian Savulescu argue together that if we are prepared to endorse the arbitrary creation and destruction of embryos in natural reproduction, we are logically committed to morally countenance cloning and other 'Frankenstein reproductive technologies' (2004). Through a close reading of a footnote in their paper, I argue that Harris and Savulescu operate with a patriarchal approach to knowledge in bioethics, and their sardonic characterisation of the wasteful creation lottery as a 'goddess' situates them in a long history of denigrating the maternal body. I then proceed to relate some of the patriarchal tradition that Harris and Savulescu uncritically inherit, which wilfully disregards or appropriates the significance of life being given by mothers in childbirth. Guided by the work of Adriana Cavarero, we look at the role of formative Greek thought in marginalising the female subject and hijacking the language of birth and parturition for the male preserve of philosophy, and show how this same operation is echoed in subsequent Christian and modern traditions of Western thought. The suggestion is that patriarchal orders have sought to denigrate the body since it is the one thing men cannot give; rather, men have valorised the gifts of their intellects with the language of birth, or else concocted second births of the spirit or rebirths that are more significant measures of existential meaning. The maternal context of the gift remains written out in the contemporary bioethical narrative that we have shifted from hapless exposure to the 'goddess' creation lottery, to newly beneficent gift-givers aided by the new technologies of reproduction.

Chapter 4 takes seriously the philosophically redacted significance of the maternal gift of birth, in light of a more deconstructive and explicitly ethical account of the gift than is afforded in standard bioethical argument. Following the established anthropological analyses of gift exchange, a recent continental tradition of thought posits the impossibility of the pure or unreciprocated gift, save that it should involve self-sacrifice and thereby extinguish the very possibility of a return. This chimes with what Grace Jantzen calls the deathly imaginary of patriarchal Western civilisation, valorising mortality and the gift of the death we can all possess as our own, over the gift of the life that is given only by women. I go on to argue in this chapter that the gift of life given by mothers is quite as radical as self-sacrifice in giving life with the impossibility of reciprocation in kind; the existentially inceptive gifts of air, time and corporeality are received in unrepeatable singularity, and can only be passed on to new, natal beings by the bodies of women. This helps to explain the patriarchal fixation upon mortality, and a history of violence and resentment against the maternal body as levelling a debt upon men that they cannot repay. The chapter goes on to develop the case for recognising the gift of life, given by maternal bodies, that also points plausibly to an account of ethical responsibility based upon a fundamental gratitude or appreciation of being given life by others.

The final chapter returns to the question of selective reproduction, reconfiguring the debate in light of a recognition of the giftedness of life from maternal bodies, considering what is lost—or what changes—with technologies that shift reproduction from delivering a maternal gift of life to unknown future persons, to a gift that determines certain conditions in advance for who those persons should be. Setting the debate in the context of Mannheim's thought on generations, and feminist correctives to Hannah Arendt's connected philosophy of natality, I suggest that the same principles guiding the intersection of generations in the activity of education apply in considering the ethics of selective reproduction. The operation of giving life unconditionally to strangers performs a function of assuring for the next generation an important freedom from the will of one's creator(s) to act without obligation to the institutions and practices of the past. The yoke of the giver over the lives they have gifted, one generation over the next, is importantly constrained in principle by the radical unfamiliarity of the child's new and unique life when they are given; our strangeness at birth secures a principled freedom from a binding obligation to those who gave us life and to refashion the world that we are given into. I therefore suggest a wider principle or biopolitics of generational beneficence to enlarge the scope of responsibility that should be built in or deferred to in the consideration of selective reproduction.

Selective Reproduction

We need hardly rehearse the fact that advances in biotechnology have changed the way that we can create human beings. In particular, conditions of infertility have been overcome for many individuals and couples desiring children who were previously unable to conceive or bear a child together naturally. This is typically cause for celebration, particularly when parents cohere with predominant social, cultural or religious ideals of familial and sexual identity. However, it's the technologies that enable the prenatal selection or prospect of 'engineering' children that tend to cause most controversy. We can adopt the broad definition of 'selective reproduction' for the choices occasioned by such technologies, following Stephen Wilkinson's description of "the attempt to create one possible future child rather than a different possible future child. The reason for wanting to practise selective reproduction is normally that one possible future child is, in some way, more desirable than the alternatives" (2010, 2). On such a definition, there is the presumption of a degree of knowledge about possible future children such that parents can select to have a particular child based on his or her desirability; it is the scope of this knowledge that has expanded so dramatically in recent years, from the first ultrasound scans, to amniotic fluid tests, to the *ex utero* technologies of Pre-implantation Genetic Diagnosis (PGD) that are possible today. The prospect of so-called gene-editing techniques also holds the future possibility of a variation on the practice of selective reproduction, such that a single embryo may be engineered to create a child with different qualities or characteristics to the child that would otherwise have developed from that same embryo.

These modern reproductive technologies mean that we need not wait until birth to learn something of a possible child's condition and can prenatally exercise selection. To take present examples, medical termination of a pregnancy may be carried out up to term in the UK when a foetus is diagnosed with a substantial risk of developing into a child who would suffer "such physical or mental abnormalities as to be seriously handicapped" (HFEA 1990). Some parents who have these terminations may do so with a view to attempting thereafter to conceive another, healthier child as a form of selective reproduction. PGD allows for earlier intervention, so that prospective parents can be assured that an embryo selected *ex utero* for implantation following IVF treatment will be free of serious hereditary conditions to which there is a known susceptibility, such as cystic fibrosis or Huntington's disease. There is a list of disorders for which PGD is licensed, which is approved and kept under review by the UK's independent fertility regulator, the Human Fertilisation and Embryology Authority.

I will take it as uncontroversial that there are harmful abnormalities which spell such suffering for a future child that exercising forms of selective reproduction can be the only humane thing to do, although as to the degree of harm or handicap that warrants avoidance or termination of a future child there is much disagreement. Down's syndrome remains a notable case in point. Looking to the future, as our knowledge of the human genome develops apace and powerful genotyping tools allow geneticists to identify the variants behind dispositions to more traits and conditions, we are faced with the possibility of ever more discriminate reproductive selections. These might be for the avoidance of genetic risk factors for more generally undesirable neuropsychiatric disorders such as schizophrenia or Tourette's syndrome, but then it may also become possible to preview and select for the likelihood of particular hair or eye colours, height, speech impediments, intelligence, musicality, sexuality, and so on.

The intellectual case in principle for extending this liberty in present and future forms of selective reproduction seems assured by a prevailing liberal eugenic current in the recent tradition of Anglo-American bioethics. While the language of eugenics is, of course, instantly emotive by association with historical state-sponsored programmes of violence and coercion, the procreative choices of prospective parents do not bear resemblance to the diabolical eugenic programmes of fascistic, authoritarian regimes. However, the practices of genetic enhancement and selective reproduction can be said to be consistent with the general purpose of Sir Francis Galton's 'science of improving stock' (1883), albeit on a more domestic scale. The concept is importantly transformed by the qualifier *liberal* eugenics, which admits on medical or libertarian grounds the connivance at offspring of choice or design, but insists upon the freedom of parents alone to make those choices with the interests of their children and families doubtless at heart. In the remainder of this chapter I will explain how a broadly liberal eugenic bearing has come to dominate and drive recent bioethical debate on enhancement and selective reproduction, specifically by way of levelling burdens of proof upon three compelling presumptions against those who object to the use of selection and enhancement technology in reproduction.

Liberty: The Democratic Presumption

The principles of liberal statehood are commonly attributed to the nineteenth century British philosopher John Stuart Mill, who radically asserted the sovereignty of the individual and the limits of state authority. Against a background of what he saw as an increasing tendency for society to impose itself upon the individual, both by force of opinion and by legislation, to compel people to adhere and defer to prescribed standards of private personal conduct, Mill defended the following principle:

That the only purpose for which power can be rightfully exercised over any member of a civilised community, against his will, is to prevent harm to others (...) The only part of the conduct of anyone, for which he is amenable to society, is that which concerns others. In the part which merely concerns himself, his independence is, of right, absolute. Over himself, over his own body and mind, the individual is sovereign. (2010, 10)

Thus, it is incumbent upon the state to ensure that society accommodates the choices of individuals so far as their actions do not harm others. An excellent, though still recent, example of the implementation of this principle was the decriminalisation of homosexuality, on the basis that the private lives of individuals are no business of the law, except when they visit demonstrable harms upon other people (Wolfenden 1957). As we know, decriminalisation by no means assures universal approval from a society, but it secures the accommodation of individuals among society with the liberty to determine and fulfil their own idea of the good. Mill identified a kind of tyranny borne beyond any straightforward political despotism in collective opinion; a social tyranny of the majority. The societal compulsion to fashion people in congruence with prevailing consensuses is as invidious a tyranny as that of any state mandate of the good life, and one from which the individual requires equal protection. Society has no more claim upon the determination of individuals than government in this respect, and must equally accommodate private diversions from popular opinion so long as they commit no harm to any others.

Mill's harm principle is commonly recognised and endorsed in liberal democracies, and is appropriated by most bioethicists as the rightful premise of moral debate on the employment of reproductive technologies. John Harris, for instance, has restyled Mill's principle the Democratic Presumption, and defines it thus:

The presumption is that citizens should be free to make their own choices in the light of their own values, whether or not these choices and values are acceptable to the majority. Only serious real and present danger either to citizens or to society is sufficient to rebut this presumption. (2007, 72)

This presumption is the confident basis for Harris's occasionally provocative approach to argument, whereby the onus is squarely levelled upon any opponent to reproductive choice, by whatever means in question, to demonstrate the harm or danger inherent in or consequent on the choice made by the individual. Note Harris's qualification that the danger must be real and present, as opposed to distant and speculative. While it might once have fallen upon technological pioneers to prove the rightness of the new or unfamiliar, the burden now ought to be the charge of traditionalists to account for denying liberty to any citizen in view of the immediate attending consequences.

A principle of Procreative Liberty speaks beyond the freedom to simply avoid conception and childbirth, to a legal prerogative over the time, manner and means one chooses positively for procreation (Robertson 1983). It falls rightly within the scope of sovereignty over oneself to select, so much as one wishes, the circumstances by which one undertakes to procreate or accedes to childbirth. This is, as Dworkin explains, a principle which follows quite plainly from the commitment to individual human dignity that underpins Western democracy; the premise that government should be "republican rather than despotic" and afford people the liberty "to confront the most fundamental questions about the meaning and value of their own lives for themselves, answering to their own consciences and convictions" (1993, 168). Undertaking to become a parent, or at least to play a part in the conception of a child, is an act through which an individual ordinarily seeks to express the values that sustain and motivate them, answering a demand to realise those values for the sake of a dependent human being. Of course, when the expression of those values plainly harms a child one becomes answerable to more than one's own conscience, but so long as no serious real and present danger attends one's procreative choices, the presumption must be that they are no business of the state.

While we would hope that procreative liberties—both to avoid having and to endeavour to have children—are relatively uncontroversial when mutually agreeable individuals entertain the prospect of parenthood through coital reproduction, a presumptive liberty right to procreate is pressed when conception proves impossible unaided or carries the foreseeable probability of creating a child with a serious genetic condition. As we know, ARTs provide or at least augur the provision of techniques to allow individuals to overcome the obstacles to their creating a healthy child. For Robertson and others, the presumptive primacy of Procreative Liberty still obtains when procreation is possible only by virtue of ARTs. That is, "non-coital reproduction should thus be constitutionally protected to the same extent as is coital reproduction, with the state having the burden of showing severe harm if the practice is restricted" (Robertson 1994, 39). Of course, the state has numberless other burdens to contend with and may not be able to provide the resources for non-coital reproduction, but it has no further warrant to intervene in private reproductive arrangements purely on account of their requiring technical assistance. Robertson makes a persuasive analogy with blind people, for whom an inability to read by sight does not controvert an interest in receiving information and therefore preclude a right to use Braille, audio recordings or a sighted reader to do so. Likewise, the interests of the coitally infertile in forming families are not qualitatively different to those of the coitally fertile and so demand equal protection as part of personal privacy or liberty. Certainly, the burden of proof, in the absence of harm, falls upon those to demonstrate such a difference, which would be hard to defend on any plausible contemporary account of justice.

The argument about sex selection in the UK demonstrates exactly the tension that arises out of a commitment to liberty when it is pressed into conflict with popular sentiment. As a result of a HFEA consultation on sex selection (2003), it is presently illegal to carry out PGD in the UK in order to select for the sex of a child based purely on parental preference. While the risk of transmitting sex-linked genetic conditions is considered good and sufficient reason to offer prospective parents sex selection, a desire to 'balance' a family, substitute a deceased child or simply entertain a particularly gendered relationship with one's offspring is not reason enough to choose its sex by embryo screening. The HFEA reported widespread public hostility to the use of sex selection for non-medical reasons, evidenced elsewhere in studies of lay people's responses to the same issue (Scully et al. 2006), and ruled that the state should continue to legislate against it. This conclusion met with fierce criticism from John Harris (2005), who rejected the authority's appeal to the speculative harm that might befall a child selected only in view of their sex. There is no empirical evidence we can call upon giving credence to this appeal, and the created child should have little cause to complain retrospectively since they would not have existed, differently sexed, but for their parents' selection.

Harris's most substantive criticism though, was on this point of presumptive liberty. He points out that the HFEA appear to commit themselves in the report to the Democratic Presumption, stating that reproduction is "an area of private life in which people are generally best left to make their own choices and in which the state should intervene only to prevent the occurrence of serious harms" (2003, para 132). The report fails to identify any decisive evidence of the likely occurrence of serious harm proceeding from non-medical sex selection, so we might then reasonably expect a (perhaps cautious) compliance with the democratic or liberal presumption that parents may exercise procreative liberty in the matter of choosing their child's sex. However, the HFEA proposed legislating against the practice in view of the recorded public hostility to the idea, concluding that "there would need to be substantial demonstrable benefits of such a policy if the state were to challenge the public consensus on this issue" (2003, para 147). As Harris observes, this seems patently to contradict the political premise of individual liberty in deference to public opinion, turning the democratic presumption on its head and subjecting harmless private lives to a formalised tyranny of the majority. Moreover, it shifts the burden of proof away from the state justifying intervention, onto the individual justifying choice, qualifying the exercise of individual liberty by the demonstration of substantial benefits. We are to accept on this matter that the state may accommodate individual procreative liberty on necessary condition of the demonstration of benefits and not simply according to a sufficient condition of harm avoidance. The burden of proof is made the individual's to bear against society, rather than the state's to honour for the protection of the individual.

The HFEA's judgement in this case appears to stand out as a glaring exception to a democratic presumption that has steered modern states to preside over mostly liberal, accommodating societies. It is this very same presumption that permits, or at least may not obstruct, British couples travelling to other countries where sex selection *is* legal in order to undertake the practice anyway. The state would require the weightiest justification for curtailing the free movement of its own people, and whilst the freedom to choose the sex of a child can hardly be regarded as of equal import, still the burden must lie with the state to account for abbreviating the liberty of its citizens. For abridging any freedom, the state should only have recourse to the consideration of harm; not religion, prejudice or prevailing public opinion. Against the burden of demonstrating such harm, there is nothing in the practice of selective reproduction that warrants state obstruction of the liberty of individuals to realise their choices.

Considerability: The Profane Presumption

Alongside the shift from individual deference to ideology to an ideological deference to the individual, there has been transference of the burden to prove moral considerability and quality of life. This burden of course is problematised particularly by the medical technologies that have allowed human beings to significantly delay, hasten, avert or manipulate natural processes at the beginnings and ends of lives. Attendant with the ascendance of the harm principle in liberal societies has been a decline in what Peter Singer (1995) has called the sanctity of life ethic. It is beginnings of lives that we are interested in here of course, and it will be relevant to account briefly for the developments that have led to our present attitudes to practices like abortion and embryo selection.

Both Plato and Aristotle advocated that the state command the killing of deformed infants (Jones 2004), so it is clear that a sanctity of life ethic has not always been a presumption of the Western conscience. According to Peter Singer, there is "no doubt that the change in European attitudes to abortion and infanticide is a product of the coming of Christianity" (2002, 229). The orthodox Christian notion that human beings are endowed with imperishable souls, coupled with equivocation over the increments of human personhood during foetal development, led the Church to abandon interpretive embryology in favour of an unambiguous threshold of moral considerability. Human life was to be regarded as sacred from the moment of conception, and its wilful destruction at any stage a manifest violation of the sanctity of that life. This doctrine took hold with the ascendancy of Christianity in Europe, and became a foundational moral presumption, eventually written into Anglo-American criminal law and even the 1948 revision of the Hippocratic oath by the World Medical Association in Geneva. Although it has since been dropped from the oath in a subsequent amendment, it remains a sincere and fundamental belief for millions of people, and a strong moral commitment in many religious traditions.

The retreat of the law from the sanctity of life ethic in considering abortion followed both from more comprehensive diagnoses of human embryonic development and viability, and constitutional commitments to individual rights of privacy and the emancipation of women. It must also be true that the role of the developed media in publicising the perils of illegal abortions and details of harmful inherited conditions stirred greater degrees of public sympathy for the political argument for legalisation. In particular, the thalidomide tragedy in the late 1950s and early 1960s, where the widely-prescribed sedative for pregnant women was linked to thousands of subsequent birth defects, helped to alter attitudes, so that "abortion no longer seemed to involve a choice between absolutes—life or not life—but matters of degree—what kind of life under what kind of conditions?" (Mohr 1978, 253). This change in emphasis also came to bear upon appreciations of euthanasia and the ethics of resuscitation and life extension for comatose patients, where that life might come to consist, by virtue of medical technology, of little more than a dependent, respiring body void of brain function. There is a gradual shift from the uncritical presumption of sanctity *qua* human life itself and a heartbeat, to a reckoning of moral value based on the quality of the life and conceivable interests of the human subject in question.

This shift has come to shape the landscape of bioethical arguments about selective reproduction and the prospect of human genetic engineering, which tend to be staged in terms of a straightforward antagonism between liberal and conservative, conditional and unconditional values. Within orthodox Anglo-American bioethics the 'conservative' sanctity of life ethic has gradually declined; the ideological triumph of liberal individualism has successfully manoeuvred moral burdens of proof onto conservatives and made them dissenters of freedoms rather than defenders of rights. The arguments of figures like Singer and Harris, which are not really different to those they were making 25 years ago, are now remarkable for being mainstream—at least in the academic literature—rather than radical as they once were (Evans and Schairer 2009, 361).

We have moved, then, from a presumption that all human life is to be valued equally at all stages, to a presumption that life's value is conditional and should be inferred from its demonstrable quality at a given time. The prevailing notion is that we have different ethical obligations to human lives at different stages in a continuum of personhood based on cognitive and physical abilities (see, for example, Koch 2006). When such a life consists in but a cluster of cells, or an inert or interminably painful body, the burden of proof is for those who would claim that there is intrinsic value in prolonging that life against the harms experienced or foreseen in its extension. A presumptive sanctity of life ethic simply does not admit of these conditions; it is usually an unconditional value, whether or not the life proceeding from its observance can or does value itself. By contrast, Harris's definition of the morally considerable being is that it be capable of valuing its own existence: "Creatures that cannot value their own existence cannot be wronged in this way, for their death deprives them of nothing that they can value" (1985, 19). One immediately virtuous implication of this definition is that it need not discriminate between species; however it becomes problematic for those human lives which have not yet begun or have ceased to value their existence. Yet if we should object by appeal to a sanctity of life ethic, it is very clear that the burden is now to demonstrate the provenance of such sacredness, and increasingly apparent that we cannot arrive at such an unconditional value without recourse to religious authority.

The UK's abortion statistics (2015) evidence plainly the accommodation and substantial acceptance of the de-sanctification of embryonic and foetal human life. Recent polling has shown that even among the 44% of people in the UK who believe that human life begins at conception, over three quarters believe that abortion is acceptable in the first twelve weeks of pregnancy, and half believe that abortion should be permitted at least up to twenty weeks (Woodhead and Winter 2013). Of course, the fact that early human life is incapable of valuing itself does not mean that it is desirable to treat such lives as of no importance; they may be valued and sustain meaning within the networks of people who bring about their existence. In this case it is the interests of these individuals, and particularly pregnant women, who have a legitimate claim to bear and welcome a child who will enter into that network or family, which may be wronged by terminating the life of a foetus. We have no claims upon or legitimate interests in the embryos or foetuses created by other people; they are not the state's to protect, and even when they might be considered sentient and susceptible to harm have not attained a degree of individual personhood warranting state veto of parental discretion in their prospective interests.

PGD is importantly different to prenatal intrauterine testing methods like amniocentesis where the technique of selection following an undesirable diagnosis is for the pregnant woman to undergo abortion. It is, I think, material to note the way in which the early screening technologies changed the experience of pregnancy for women. De-sanctifying foetal life was the conceptual moral change; the technologies of pregnancy, as Rothman famously observed, made the experiential shift for women from complete and immediate attachment to a guarded separation from the foetus. The conditional value we came to vest in the foetus was reflected in the conditional maternal attachment encouraged by these technologies, which could only diagnose the foetal condition halfway through the pregnancy. Thus women became directed to perform what Rothman (1993) calls the tentative pregnancy, geared to recognising the foetus as separate from the mother and employing technology with the assumption that mothers "start from separation and come to intimacy—and only with caution" (115). Likewise, in our regard for the status of human life, we now start from separation and come to moral obligation too with some caution.

Elsewhere, geneticist Angus Clarke has perceptively criticised the putative non-directiveness of the counselling offered to prospective parents in anticipation of prenatal screening. He contends that "an offer of prenatal diagnosis implies a recommendation to accept that offer, which in turn entails a tacit recommendation to terminate a pregnancy if it is found to show any abnormality" (1991, 1000). That is, acquiescence to prenatal diagnosis implies assent to the clinical logic of the tentative pregnancy, affirmed and approved by degrees. Moreover it is the tentative pregnancy that is the prepared and practised choice regime for the mother, the pregnancy around which the professional lives of the attending medical personnel are structured and rehearsed. The tentative pregnancy is in fact the norm, it is the mature and responsible approach to retaining charge over one's own life and the quality of life of one's child to come. There is a greater onus to explain refusal than assent to the offer.

Striking in both Rothman's account and Clarke's criticism of reproductive technologies in practice, is the subtle shift in burdens eventuated by the abandonment of the sanctity of life ethic. A presumption of the sacredness of human life levels the burden of proof very much at those who would abort it, to demonstrate that a life would not, after all, be of sufficient quality to be worth living. By contrast, what I am tempted to call, in the spirit of John Harris, the Profane Presumption at work in secular bioethics and reproductive healthcare, tilts this burden at parents to prove that their child's life *will* be worth living. With this presumption at play, the value is obviously in seeking to find out and, if necessary, selecting for a child with demonstrably higher chances of leading a life of considerable quality or less harm than the alternatives. While PGD enables mothers to avoid the damaging, detached experience of the tentative pregnancy and potential harm of an abortion, it is an extension of the same logic that makes for tentative *in vitro* conceptions instead. Indeed, when we see this logic through we arrive at what Julian Savulescu has called a principle of Procreative Beneficence: the moral obligation, where selection is possible, to create children with the best chance of the best life (2001). I will return to this principle in the chapters that follow.

Nurture: The Parity Presumption

Regard for human life as sacred from conception entails social and medical commitments to improve the lot of those who are born with undesirable traits or inherited conditions. Of course, this is also true of those who may define valuable life in more contingent terms, however the technologies of selective reproduction now afford parents an option of prevention that might not be countenanced under strict ethics of human sanctity. A policy of prevention resounds with charges of eugenics, tending towards concerted and deliberate efforts to eliminate undesirable conditions rather than commit to treating or accommodating the future individuals who would possess them. This objective is recognisable in the reasoning behind the most unspeakable crimes of the last century, and the haunting phrase *Lebensunwertes Leben*: life unworthy of life. Yet this science has been reclaimed by a liberal orthodoxy in bioethics, which "adopts a more or less aggressively eugenic face in its axiomatic assertion of the conditional nature of the valued being" (Koch 2006: 256–257). The difference is that the condition of a human being may be divined *before* it is a valued or self-valuing being, so the injury of its destruction, though in the name of the same science, is sidestepped with the devolution of choice to individuals. For advocates of the new liberal eugenics, this unlocks a parity claim between interventions in nature and nurture, the burden of disproving which is levelled at their critics.

The case of reform or liberal eugenicists often involves arguing from precedent, averring that uses of reproductive technologies to particular ends are in fact consistent with practices we already regard as permissible, or are at least different in no morally relevant way. The moral invulnerability of the embryo occasions the claim for the parity of genetic and environmental interventions. Critics of genetic modification or selection need to advance a morally relevant difference "between moulding a child after its birth by the manner in which one rears it and moulding a prospective child before it is born by genetic means" (Archard 2007, 6). Clearly we are permitted to shape and fashion our children after birth by way of education, enculturation, health and welfare provision; we do not attribute so noble a purpose in nature as to preclude our vaccinating against disease or seeking to remediate genetic disadvantages. We create and select environments and cultures for children which promote their flourishing; we avoid environments that would do them harm. In arguing from the precedent or 'moral image' of nurture, Nicholas Agar contends that "If we are permitted to produce certain traits by modifying our children's environments then we are also permitted to produce them by modifying their genomes" (2004, 113). We should also add, in view of the practice of PGD, that this also holds good for genome selection as well as modification: if we are permitted to select for children environments or activities conducing to certain traits we ought also to be permitted to select children with the genetic predisposition to those traits.

Any substantive objection to selective reproduction is thus subject to the burden of demonstrating how it is different in a morally relevant or more serious way to the conditioning, nurturing and educational practices we already permit or encourage. That they may fail to demonstrate such a difference, it should be noted, does not establish by itself the rightness of selective reproduction; we should allow that our considerations of moral parity could give cause to reflect unfavourably upon present policies and attitudes. If we take the charge of commodification, for instance, we should hardly be able to claim that parental selection of embryos using PGD is *prima facie* more objectionable than the commodification of some naturally conceived children. Whilst it is rare that children are created solely as instruments for some ulterior parental purpose, it is common that parents have motives for procreating which tend to instrumentalise the child, such as improving a marriage or relationship, pleasing other family members, providing an heir, securing additional state benefits, etc. However, we might not consider them especially blameworthy provided the child is also viewed as a human being, valued in her own right and, to use Kant's idiom, treated as an end-in-herself. The objection of commodification is faced down with the following burdens: "Can they show that the reproductive practice under discussion involves not just treating children as means, but treating them only as means? And can they show that the degree and kind of instrumentalisation is worse than that involved in many very normal and widely accepted scenarios?" (Wilkinson 2010, 135). Given the incidence and circumstances of PGD at present, we should concede that the degree and kind of instrumentalisation involved appears no more objectionable than practices we already allow.

It is not a wholly convincing form of moral argument to assert that a thing is certainly no worse than we allow already and must therefore be permissible; it may be that 'very normal and widely accepted scenarios' are far from ideal. This seems true of what can often be allowed by way of parental influence over a child's autonomy, which is the concern of one of the more coherent challenges to procreative liberty through PGD advanced by Jürgen Habermas. Autonomy is a vexed philosophical and political concept, but for our purposes we need not depart from Feinberg's clarification in the specific case of children:

> When the state justifies its interference with parental liberty by reference to the eventual autonomy of the protected child, it argues that the mature adult that the child will become, like all free citizens, has a right of self-determination, and that that right is violated in advance if certain crucial and irrevocable decisions determining the course of his life are made by anyone else before he has the capacity of self-determination himself. (1992, 91)

Of course, families have always presented problems for individual self-determination, and we need not quote Philip Larkin to be reminded that parental influence can be lasting and regrettable. Individuals do not graduate from childhood with a pure capacity of self-determination intact, and perhaps we could hardly conceive of such a thing; nevertheless it is surely true that this capacity can be more or less impaired for individuals by the actions of their parents, and it is at least highly plausible that the least or lesser impairment should be morally preferable.

For Habermas, the critical aspect of the violation visited upon a child's autonomy through PGD is that its decision determines irrevocably the course of the child's life. The violations usually occasioned in the course of a child's upbringing are, if not reversible, subject to the revision of the individual when they become capable of self-determination. Thus, a child raised as a carnivore may eventually reach a mature decision to cease eating other creatures: of course she cannot un-eat the meat she has eaten but may console herself with the fact that her childhood environment and diet was not her own to determine. A child may even be beset by neurotic fixations as a result of some chronic neglect or abuse by their parents, but, according to Habermas, could still "retrospectively compensate for the asymmetry of filial dependency by liberating themselves through a critical reappraisal of the genesis of such restrictive socialisation processes" (2003, 62). The particular problem with PGD and genetic engineering is that since the parental decision is constitutive of the child's genome she can never be liberated of it to become the undivided author of her own life. The intentions of a child's parents are incarnate in her very makeup so her "own body presents an expectation she can't escape" (Prusak 2005, 36). Environmental influences, or nurture, may be challenged or revised whereas deliberately contrived genetic influences, or nature, may not: for Habermas this decisively undermines the claim to a moral parity of influences tended through genes and environment, nature and nurture.

Habermas appears to have identified an important qualitative difference here, and a violation exceeding those to which a child's autonomy should normally be exposed. His contention is that there is a moral quality to the contingency of our origins, which is necessary for a person to be capable of regarding herself as an autonomous or self-governing being. We all, no doubt, are pitted against parental expectations, but to have such expectations deliberately stitched into our genes may leave us railing against our own bodies for 'authorship' of our lives. This sounds quite compelling in the abstract, but less convincing when we consider it against the complex reality of self-determination. After all, some of the more severe or extreme demonstrations of parental neglect and maltreatment confound Habermas's dubious prescription that individuals might simply be able to reappraise their socialisation in order to be liberated of childhood abuse. Early environmental influences may be as irreversible as genetic influences, or indeed more so, as it may transpire in the future that people can reassign or reverse the genetic qualities for which they were selected. In the meantime, a surely graver battle with one's body is fought by those in a daily struggle against debilitating inherited diseases rather than the expectations of one's parents, which after all still rely upon the individual's cooperation for their expression.

If anything, we might say Habermas's argument vindicates the orthodox view and practice of PGD, which tends to be applied therapeutically and indeed under the pretext of the child's autonomy. When the HFEA licenses PGD for serious genetic conditions it is, in part, because those conditions may gravely limit a future person's capacity for self-determination in ways that cannot be compensated for or overcome by environmental modifications. There is a case made here by disability activists that some of these judgements are thus motivated by contingent and derivative norms of ability (see, for example, Shakespeare 1998). Without the space to account adequately for these arguments here, I shall take it that some individuals are yet embodied in ways that cannot afford them reasonable capacities of self-determination and do indeed augur lives of immense suffering. The liberal eugenicist is less likely to argue that procreative liberty should extend to a right to deliberately select for a child possessing such a condition, and it is hard to conceive that a parent should want to do this. Rather, the ethics of bearing children might be guided by the same principles we observe in the rearing of them. These we might plausibly regard as promoting what Feinberg (1992) called the open future of the child. So just as parental freedom is limited such that the child may not be exposed to environmental influences that restrict her future autonomy, so procreative liberty might also perhaps be bound in principle to the same constraints by way of selecting a child's genetic makeup.

A commitment to resolve unmerited misfortune is not only an impulse of conscience but also a fundament of the society we inhabit in observance of the Democratic Presumption. The state is licensed to limit individual liberty when its pursuance visits harm upon other individuals, and so we might say that such harm—including serious violations of future autonomy—is unjust. Not only does the state assume powers of restorative and punitive justice after the commission of harm, but it is also sanctioned to act in advance of harm with sufficient cause by refusing harmful individuals certain social privileges such as rearing and educating children. The state does not, in principle at least, admit of an environmental lottery in the distribution of social goods like education, so the moral parity of influences tended by nature and nurture suggests that there may be equal reason to insure against brute bad genetic luck in the interests of justice. Buchanan et al. affirm genetic technologies and advocate, on the basis of the parity presumption, a moral case for the colonisation of the natural by the just: "the domain of justice extends in principle to natural as well as social assets" (2000, 102). Through provision and license of ARTs the state supports prospective parents who endorse this view and exercise a liberal eugenic discretion in the selection of children free from very gravely unfortunate conditions. And so long as the scope of other selections is consistent with benign parental aspirations, it must be shown that these are demonstrably more harmful as a matter of prenatal than postnatal choice.

I've sketched the background to what has developed as an orthodox liberal eugenic view of selective reproduction in philosophical bioethics. This view is borne out in practice where the offer of prenatal screening and therapeutic abortion is routine for high risk factor mothers, and PGD is licensed in the UK to enable parents to avoid the inheritance of serious genetic conditions for their children. The orthodox bioethical position is that this is morally sound in principle, since it accords with presumptions of individual liberty and the conditional value of human lives. Moreover, in the absence of a compelling morally relevant difference between eugenic and environmental influences, purposive selection for genes, or 'nature' as well as nurture may be regarded as permissible or even virtuous parental action, for therapeutic and enhancement purposes or for any other benign parental preference. This admittedly brisk summary of the arguments for liberal eugenics, describes the bioethical burden that has been brought to bear on those yet uneasy or opposed to the technologies of choosing children; in the following chapters I try to pick out a different narrative about the use of these technologies, beginning with an argument from gifts.

References

Agar, Nicholas. *Liberal Eugenics: In Defence of Human Enhancement*. Malden, Mass; Oxford: Blackwell Publishing, 2004.
Archard, David. "Genetic Enhancement and Procreative Autonomy." *Studies in Ethics, Law, and Technology* 1, no. 1, (2007).
Buchanan, Allen E., Dan W. Brock, Norman Daniels, and Daniel Wikler. *From Chance to Choice: Genetics and Justice*. Cambridge, U.K.: Cambridge University Press, 2000.
Chief Medical Officers. Abortion Statistics, England & Wales: 2015. edited by Department of Health. https://www.gov.uk/government/uploads/system/

uploads/attachment_data/file/570040/Updated_Abortion_Statistics_2015. pdf 2015.

Clarke, A. "Is Non-Directive Genetic Counselling Possible?". *Lancet* 338, no. 8773 (1991): 998.

Dworkin, Ronald. *Life's Dominion: An Argument About Abortion and Euthanasia*. London: HarperCollins, 1993.

Evans, John, and Cynthia Schairer. "Bioethics and Human Genetic Engineering." In *Handbook of Genetics and Society: Mapping the New Genomic Era*, edited by Paul Atkinson, Peter Glaser and Margaret Lock. Oxford: Routledge, 2009.

Feinberg, Joel. *Freedom and Fulfillment: Philosophical Essays*. Princeton, N.J.: Princeton University Press, 1992.

Galton, Francis Sir. *Inquiries into Human Faculty and Its Development*. [S.l.]: J. M. Dent and Co, 1883.

Goleniowski, Hayley. Downs Side Up. http://www.downssideup.com/ Accessed 9th March 2017.

Habermas, Jurgen. *The Future of Human Nature*. Cambridge: Polity, 2003.

Harris, John. "Sex Selection and Regulated Hatred." *Journal of Medical Ethics* 31, no. 5 (May 2005): 291–294.

———. *Enhancing Evolution: The Ethical Case for Making Better People*. Princeton, N.J.: Princeton University Press, 2007.

———. *The Value of Life*. London: Routledge & Kegan Paul, 1985.

———. "Human Fertilisation and Embryology Act." http://www.legislation. gov.uk/ukpga/1990/37/pdfs/ukpga_19900037_en.pdf 1990.

———. "Sex Selection: Options for Regulation. A Report on the HFEA's 2002–2003 Review of Sex Selection Including Discussion of Legislative and Regulatory Options." http://www.hfea.gov.uk/docs/Final_sex_selection_ main_report.pdf 2003.

Jacob, Jennifer. "(Unexpected) Stories of a Down Syndrome Diagnosis." http://www.missiont21.com/prenatal. Accessed 9th March 2017.

Jones, David Albert. *The Soul of the Embryo: An Enquiry into the Status of the Human Embryo in the Christian Tradition*. London: Continuum, 2004.

Koch, T. "Bioethics as Ideology: Conditional and Unconditional Values." *Journal of Medicine and Philosophy* 31, no. 3 (Jun 2006): 251–267.

Mill, John Stuart. *On Liberty*. London: Penguin, 2010.

Miller, Barara. "Down Syndrome: Parents Say They Feel Pressure to Terminate Pregnancy after Diagnosis." ABC News, http://www.abc.net.au/ news/2016-11-22/down-syndrome-parents-pressured-to-terminate-preg-nancy/8033216. Accessed 9th March 2017.

Mohr, James Crail. *Abortion in America: The Origins and Evolution of National Policy*, 1800–1900. New York: Oxford University Press, 1978.

Oliver, Kelly. *Technologies of Life and Death—from Cloning to Capital Punishment*. New York: Fordham University Press, 2013.

Prusak, Bernard G. "Rethinking "Liberal Eugenics."" *Hastings Center Report* 35, no. 6 (2005): 31–42.

Rapp, Rayna. *Testing Women, Testing the Fetus: The Social Impact of Amniocentesis in America*. New York; London: Routledge, 1999.

Richards, Clare. *A World without Down's Syndrome?* BBC Television, 2016.

Robertson, J. A. "Procreative Liberty and the Control of Conception, Pregnancy and Childbirth." *Virginia Law Review* 69, no. 3 (1983): 405–464.

Robertson, John A. *Children of Choice: Freedom and the New Reproductive Technologies*. Princeton: Princeton University Press, 1994.

Rothman, Barbara Katz. *The Tentative Pregnancy: How Amniocentesis Changes the Experience of Motherhood*. New York: Norton, 1993.

Savulescu, J. "Procreative Beneficence: Why We Should Select the Best Children." *Bioethics* 15, no. 5–6 (2001): 413–426.

Scully, J. L., T. Shakespeare, and S. Banks. "Gift Not Commodity? Lay People Deliberating Social Sex Selection." *Sociology of Health & Illness* 28, no. 6 (Sep 2006): 749–767.

Shakespeare, Tom. "Choices and Rights: Eugenics, Genetics and Disability Equality." *Disability & Society* 13, no. 5 (1998): 665–681.

Singer, Peter. *Rethinking Life & Death: The Collapse of Our Traditional Ethics*. Oxford: Oxford University Press, 1995.

Singer, Peter, and Helga Kuhse. *Unsanctifying Human Life: Essays on Ethics*. Oxford: Blackwell, 2002.

Solomon, Andrew author. *Far from the Tree: A Dozen Kinds of Love*. New York: Simon and Schuster, 2012.

Stewart, Rory. "What is the opium of the people?" *Intelligent Life*, November/December 2013.

Wilkinson, Stephen. *Choosing Tomorrow's Children: The Ethics of Selective Reproduction*. Oxford: Oxford University Press, 2010.

Wolfenden, John Frederick Sir. *Report of the Committee on Homosexual Offences and Prostitution*. [S.l.]: H.M.S.O., 1957.

Woodhead, Linda, and Norman Winter, eds. *Religion and Personal Life*. London: Darton, Longman and Todd, 2013.

2

Gift and Beneficence

Not only do most cultures classify human life as a gift. But they take in particular the life of a newborn child to be a gift that has been bestowed upon its parents.
(Hyde 2007, 97)

Abstract Resistance to procreative liberty is often framed in language about life's giftedness, and the impropriety of choosing gifts in the persons of one's children. This chapter examines Michael Sandel's treatment of this argument and further flawed attempts to distinguish the gift from the merely given. Against the idea of children as gifts, antinatalism contends that life is never so good as to justify being given, and Procreative Beneficence that we ought to choose the best possible lives. This chapter argues that both positions retain an ethic of giftedness, but instead figures parents as giving rather than receiving gifts in their children, where in the case of Procreative Beneficence the capacity to give life well follows from the purposive selections enabled by assisted reproductive technologies.

Keywords Gift · Procreative beneficence · Children
Chance · Anti-natalism · Reproduction

© The Author(s) 2017
S. Reader, *The Ethics of Choosing Children*, Palgrave Studies in Ethics and Public Policy, DOI 10.1007/978-3-319-59864-2_2

In many cultures and traditions, it's common to find life referred to or understood as a gift. But does this notion get us anywhere beyond, perhaps, expressing a healthy appreciation for existential good fortune or an unfashionable religious conviction? Can it make sense to think of children as gifts bestowed upon their parents if we're unable to conceive of an intelligible agency behind the bestowal? As Lewis Hyde observes, such a conception of life is undoubtedly a part of our cultural inheritance, yet advances in scientific understanding, along with new forms of procreation and family life, make it unclear as to whether the language of giftedness tells us anything useful about parenthood and reproductive ethics.

Most conventionally, of course, we find claims for life's giftedness proceeding from religious perspectives that credit a deity with giving life. On such a view, any wilful prevention of a human birth during or following the process of conception, can represent a moral affront against God's creation. In the HFEA's consultation on sex selection for non-medical reasons, respondents were noted as objecting that "because children are the 'gift of God' parents should not seek to choose the kind of children they will have but should gratefully accept and nurture whatever children they have" (2003, para 90). Of course, the small matter of testing this proposition is somewhat complicated by grander metaphysical questions than can be addressed here. But if one rejects the notion of a divine giver, or at least supposes that the burden of existential proof for such a deity has not been met, the giftedness invoked from a personal God is contingent upon a relationship that is still evident only to the faithful.

Giftedness, however, remains a theme that is mobilised in discussions about the morality of selective reproduction without resort to God, as evidenced by a study into the 'ordinary ethics' of lay people considering the morality of social sex selection for non-medical reasons (Scully et al. 2006). In a 2002 UK study, a scenario was presented for group discussion whereby a couple with three daughters planned to have another child and wanted to use PGD in order to ensure they had a son. While there is a broadly permissive view of such action on the basis of individual freedom in what the authors call "secular-liberal bioethics," the study found that 83% of lay participants in group discussions held that it was *not* morally justifiable. Within these discussions, the idea that children are a gift was one that stood out: participants tended to speak of children as gifts and of parents properly receiving them as such. Since scant reference was made to God as giver of the child's life, the claims were understood as being a metaphor, "used to convey something important about how the speaker believes parents should relate to their children, and the responsibilities this relationship involves" (753). Drawing on the customary etiquette of gift-giving in Western European culture, it was observed that such a metaphor implied the preferred virtue of acceptance; that is, parents ought to accept children just as they come, without rejecting, selecting or seeking to change their characteristics.

Within the established secular literature on reproductive bioethics, the most noted attempt to flesh out this idea comes from American philosopher Michael Sandel. In the first part of this chapter I will attend in more detail to his arguments from giftedness against genetic enhancement and selective reproduction, along with attempts to refine and substantiate them. Ultimately, Sandel's philosophical exposition of what are characterised as the intuitive, emotive—or even hateful (Harris 2005)—responses of lay people to selective reproduction is found to be generally unpersuasive, even when developed further and revised by other thinkers. However, the discussion of giftedness in reproduction precipitated by Sandel's arguments does open up a line of thought that identifies some of the same themes in precisely those thinkers that oppose him. While Sandel is concerned with the morality of receiving the gifts of life, an ethic of giving life well is central to the principle of procreative beneficence expounded by Julian Savulescu, who pits his argument directly against Sandel, and to the anti-natalist position of David Benatar, for whom life is never so good a thing as to warrant being given. Sandel, echoing wide public opinion, concentrates upon the etiquette of how life ought to be received as a gift, but the agency facilitated by reproductive technologies to select or create children by design switches the focus for their advocates onto the proper virtues of how life ought to be given. By drawing out this narrative shift in the ethics of reproduction aided by new technologies, which depict these as privileging parents anew with the capacity to give life well to their children, we will also begin to see how this account colludes with an historical effacement of maternal giving that will be taken up in the following chapters.

Sandel and the Gift of Children

At the heart of many critiques of Sandel's *The Case Against Perfection* (2007), often dismissed as just so much rhetoric, is a fundamental disagreement about the language proper to the debate. Sandel is clear at the beginning about what he considers the failure of mainstream bioethics to capture what is at issue:

> In liberal societies, they reach first for the language of autonomy, fairness, and individual rights. But this part of our moral vocabulary does not equip us to address the hardest questions posed by cloning, designer children and genetic engineering … about the proper stance of human beings toward the given world. (9)

Sandel reaches for a different kind of language, excluded from the limited moral vocabulary of liberal societies where expressions of our relation to a 'given world' have been crowded out. We're now suspicious of such expressions lest they hide an agenda to revive a recourse to some theology or ideology that would curtail our hard-won individual freedoms. It is increasingly difficult to argue from any first principles or accounts of a 'given world' to a positon that collides with the presumption of individual liberty, particularly with respect to sovereignty over one's own body. And rightly so. But it ought to be legitimate to stake out the shape of the moral terrain that might follow from the new liberties we look to enjoy, and their effect upon a world that is unaccustomed to them.

While Sandel echoes Habermas's concern for the autonomy of eugenically selected children, he asserts that the case against selection cannot rest on liberal grounds alone. Rather, Sandel argues that "eugenic parenting is objectionable because it expresses and entrenches a wrongful stance toward the world—a stance of mastery and domination that fails to appreciate the gifted character of human powers and achievements" (83). An appreciation for such giftedness is vital to our moral landscape according to Sandel, and finds its quintessential expression in parental openness to the unbidden characters and natures of our children. "To appreciate children as gifts is to accept them as they come, not as objects of our design, or instruments of our ambition" (45). Thus attempts to engineer in one's children certain traits or qualities of one's choosing are acts of hubris and moral hazard that corrupt the norms of parenting and diminish the parent. It is not just, then, about preserving the autonomy of one's children, but also preserving certain norms of parenthood that serve not just as instruments to childhood flourishing but also to morality in general.

Specifically, Sandel suggests that eugenic parenting, as it diminishes our appreciation for the gifted character of human qualities, will transform three important features of our moral landscape—humility, responsibility and solidarity. He argues that a recognition of being born with talents and abilities that are neither appointed or designed but rather given, restrains a tendency toward hubris and informs a sympathy for the similarly given fortunes and misfortunes of others. As such, our humility may give way with the technologies of procreative choice; we begin to attribute less to chance and regard parents as "responsible for choosing, or failing to choose, the right traits for their children" (87). For some parents, this may not appear especially regrettable, but Sandel points to the considerable moral burden this begins to place upon us. Where prospective parents cannot avoid making choices in the enlarged frame of moral responsibility that accompanies new habits of reproductive control, there are further daunting implications to the consequences of their decisions. "When genetic screening becomes a routine part of pregnancy, parents who eschew it are regarded as "flying blind" and are held responsible for whatever genetic defect befalls their child" (89). We can speculate further that such responsibility could extend to the financial burden involved with the care and treatment of a child with a serious genetic condition since it was within the power of the parents to avoid it; society may begin to question why anyone else should be answerable for the costs of such care when the condition—or at least the risk of it—is knowingly allowed by the parents. Sandel observes that many parents of children with Down's syndrome or other genetic disabilities feel increasingly judged or blamed because they could have chosen otherwise. An appreciation and preservation of the giftedness of children could help resist a contraction of our social and moral responsibility for the lives of others beyond our own kin.

Sandel warns that the expanded reproductive choices and thereby inferred responsibility for both one's own fate and that of one's children "may diminish our sense of solidarity with those less fortunate than ourselves" (89). An appreciation of the giftedness of the natural talents that allow us to flourish keeps us alive to a duty to those who lack comparable gifts. For Sandel, the question of why we owe anything to the least advantaged members of society is answered compellingly by the appreciation of our existential good fortune; with a declining sense of the contingency of our gifts in a world where they are appointed by and for us, we may slide "into the smug assumption that success is the crown of virtue, that the rich are rich because they are more deserving than the poor" (91). There is a connection, then, between an appreciation for giftedness and the moral sense of solidarity with those others who have been exposed to and created by an equally contingent accident of fate in conception, which simply and indiscriminately furnishes some people with more advantages than others.[1] For Sandel, the foreseeable consequences of enhancement and selective reproduction upon the essential moral features of humility, responsibility and solidarity establish a significant reason to resist such technologies.

The problem that Sandel remains answerable for is drawing lines around which gifts or given qualities are to be honoured and for what reason; which interventions in a 'given world' betray a drive to mastery and which are legitimate undertakings. If there is a moral aspect to preserving a passive relation to reproductive contingency, how passive ought we to be in relation to a wide spectrum of given harms, misfortunes and disadvantages? And if there is only chance in the contingency of human origins, what reason can we have to resist the preferences of individuals to select or engineer against harmful or undesirable conditions? It may be proper to retain a sense of giftedness as an aspect of good character with regard to the domain of contingency so far as that domain extends, but where that domain may be annexed by human control an appeal to giftedness should not arrest the contrivance of enhanced or more preferable outcomes than chance would afford.

Let's take an example, and suppose that one were staging a cricket match in a changeable climate and happened to have a clement day; it may be apt to consider the sunshine a gift or as something for which one ought to be grateful. That is to say, one had neither the means nor the right to command or expect the sunshine, so an appreciation of one's good fortune would be consistent with good character. However, if one were availed of the latest weather modification technology and employed cloud dispersal chemicals thereby ensuring that play would not be rained off, it would be inappropriate to behold the dry weather as a gift since one had taken measures to guarantee it. One may yet in good character retain a sense of giftedness for the technology and the inventiveness of its creators, but it would not be correct to regard the dry weather itself as a gift since its origins were not a matter of chance. And the presumption should of course be against preserving a passive relation to the contingency of the weather in light of the clear benefits of securing an uninterrupted day of cricket.

Likewise, where the technologies of reproduction begin to crowd out the contingencies that bring about harms or disadvantages for our children, the presumption must surely be in favour of a parental liberty to be availed of them over any abstract valorisation of contingency itself. It's certainly proper to be appreciative of serendipity where desirable outcomes arise by chance, but where such outcomes may be effected by design the preservation of a sense of giftedness should not become a prohibitive end in itself. Few people would argue that we should fail to remedy embodied disadvantages or deteriorations such as sight or hearing loss to honour the vagaries of genetic misfortune. By the same token then, the preservation of this sentiment, valuable as it may be, should not be a barrier to making any practicable selections and enhancements that provide for lives better designed to go well in the world.

Gifts, Givens and Goods

Sandel himself acknowledges that his argument may appear overly or inescapably religious and might seem, therefore, to stand or fall on the plausibility of theism itself (92). Arthur Caplan, himself a critic of enhancement, has reflected that "the gift makes no sense in the secular context such as Sandel proposes. Gifts require a giver but nature offers no likely suspects to occupy this role" (2009, 208). While Sandel generally restricts himself to speaking of a *sense* of giftedness, or the gifted *character* of human powers and achievements, the philosopher Michael Hauskeller has mounted a defence of Sandel, undertaking to distinguish the gift from the merely given in order to demonstrate the moral significance of resisting genetic enhancement. He takes a common criticism of Sandel, voiced by Leon Kass, who points out: "The mere giftedness of things cannot tell us which gifts are to be accepted as is, which are to be improved through use or training, which are to be housebroken through self-command or medication, and which opposed like the plague" (2003, 19). Hauskeller then endeavours to distinguish between the gifted and the given to enable us to do just that.

The first of Hauskeller's requirements is that "a gift is something that has been given to us as a *good*. If *someone* has given it to us, then it must at least have been *intended* to benefit us" (2011, 62). Doubtless most gifts are indeed given in this spirit, and we might recognise, even in receipt of useless presents, that the giver's heart was in the right place. Yet there is notable ambiguity about whether the good is located in the given thing itself or only in the intention of the giver. Hauskeller continues: "If you give me a load of rubbish, I will hardly consider it a gift, even if I get it for my birthday (unless, of course, I like rubbish, but then it is no longer rubbish to me)" (62). It is unclear exactly what Hauskeller means here by 'a load of rubbish,' but his example raises the problem of judging giftedness when the intentions of the giver are not met by the recipient's attitude to the thing given; when the gift is not *received* as a good. He seems to allow that, in the unlikely event of the recipient being a liker of rubbish, to be given a load of rubbish for one's birthday may then be considered a gift. But there are other scenarios here if we are to allow that the value of the thing given depends upon the estimation of the agents involved. For instance, I may in good faith give you a load of rubbish for your birthday, perhaps believing that you share *my* liking for rubbish and intending it as a good, but it may be received as a nuisance or insult. It is certainly given as a good, but not received as such, yet it seems to meet Hauskeller's definition for giftedness. It may also be that, although I know that you like rubbish, I consider it an unhealthy and corrupting inclination; if I yet give you a load of rubbish knowing that you will receive it as a good while I consider it harmful and unwholesome, the giftedness of the rubbish is muddier still. I might instead decide that a greater good to give you would be a self-help book, designed to cure you of your penchant for trash, but you may consider this offensive or paternalistic.

Hauskeller invokes his distinction in order that we can recognise things that ought to be accepted as gifts, and things merely given that may therefore be opposed, 'housebroken' or enhanced. Proceeding from the questionable principle that a gift must be something that has been given as a good then, he moves without argument to claim that where we receive a thing from no one in particular, "I can then still regard it as a gift, but only if it really is good or at least appears good to me" (62). Accordingly, he suggests, for example, that proneness to disease may be given but is not rightly considered a gift, while a healthy body *can* be thought of as a gift because it is both given *and* a good thing to have. Yet without a coherent or identifiable giver, and therefore any intent in what is given, we appear to be left with little more than contestable claims about what appears as good or *good enough* to the recipient; if there is no giver, or good in mind with what is given, why should we harbour any moral qualms about rejecting or enhancing what we receive, much less characterise only that which appears good to us as morally significant? The language of giftedness seems superfluous here, or if one wishes to retain it to assign a morally compelling aspect to the goods we receive, it would be no less arbitrary to suggest that the harms we are given are to be accepted by the same token. Hauskeller's distinction augurs the possibility of drawing a line between treatment and enhancement, resisting the latter to conserve and honour our 'gifted' goods while submitting given ills to the remediation they deserve. However, without the intent of a giver, this approach only reduces to a kind of emotivism about what some people consider *good enough*, drafting in the rhetoric of giftedness to forestall any further goods we might want to confer upon ourselves or others.

Hauskeller wants to attach conditions to things to help us differentiate between the gifted and the merely given; that is, the morally significant from the incidental. He appears to claim that, ordinarily, givenness is a necessary though not sufficient condition of a gift; it also requires good intention on the part of the giver. Yet he then also seeks to claim that there are natural gifts for which givenness, in the sense of a purposive gesture, is not a necessary condition, and for these the mere recognition or interpretation of goodness alone is sufficient. That is, where givenness—and with it good intent or beneficence—is not apparent or indeed is inconceivable, the perception of goodness or good fortune in having a thing is quite enough to attach the same morally binding or gifted quality to it. Goodness becomes a sufficient condition alone for giftedness, and a gift is only that which 'really is good or at least appears good to me.' In other words, if I consider an un-given or accidental quality good, I may say that it is a gift and, by that token, it ought to be honoured as such. Hauskeller's turn of phrase here is interesting; his reference to that which '*really* is good' seems to acknowledge that there is some dispute over the nature of goodness, or perhaps that there is some objective quality of the good which can be confused or corrupted. In any case, without the discernment of intent from a giver, the estimation of goodness falls upon those who possess or are affected by whatever it is said has been given. And as we know, such estimations differ. Person A may consider the sunshine a gift as it allows for a full day of cricket; person B may consider a thunderstorm a gift at precisely the same time and precisely because it prevents it. While these may be trivial preferences, it illustrates the fact that what is considered good by one person can be flatly contradicted by what is considered good by another; it would not be sufficient argument against bringing the covers on at the cricket for person B to claim that *they* considered the thunderstorm a gift.

Perhaps the most critical ethical point of distinction for Hauskeller, between the gifted and the merely given, is that "a gift is not a loan. A loan has to be returned to the lender. A gift, on the other hand, has to be accepted and kept … One has to take good care of the gifts one has received, cherish them, even if one does not like them very much" (63–4). This, of course, is the central principle behind the moral case against choosing one's children; the gift of one's own life and the lives of one's children should be received well and accepted rather than opposed or modified. Yet note Hauskeller's own concession that one might not very much like the gifts one receives; that is to say, presumably, that one might not consider such gifts a good even if they were given as such.

He is again locating giftedness solely in the good intent of the giver, acknowledging that what is given may not actually be a good in itself, but so long as it is given as a good we must honour the gift by preserving it. This is not an immediately compelling claim in any case, for if something given as a good is, in fact, so disagreeable as to be harmful, the notion that one is yet obliged to cherish it is certainly questionable. But in the matter of our natural gifts of life and health, things that we think have not been given to us by anyone in particular, there is no good intent for us to honour but, again, only the question of whether, in Hauskeller's words, the gift 'really is good or at least appears good to me.' If the gift does not appear good—or good enough—then, and there is no giver or good intent against which one's failure to cherish the gift might offend, it is unclear why one is yet obliged to accept what has or may be given. The only argument for Hauskeller's claim that gifts have to be accepted and kept, is that they "can only be returned at the risk of insulting" the giver (64). Quite apart from the question of whether an insult is tantamount to a moral offence, if there is no agency in the giving there can surely be no such risk anyway.

Hauskeller's attempt to refine Sandel's objection to enhancement by distilling the concept of gifts from the merely given, illustrates the difficulties of trying to draw legitimate moral conclusions from such a distinction. He needs to establish that there are two kinds of gifts: those from distinct and identifiable givers, and impersonal or natural gifts from no one in particular. Yet since he says a defining criterion of the gift is that it is given to us as a good, he must do more to insist that the second kind of gift remains properly a gift when we cannot discern any good intent in the given. He seems to provide a definitive feature of gifts, acknowledge an exception in what we appear to be given from no one in particular, yet maintain that these too are gifts simply by virtue of our believing them to be goods that we have no legitimate claim to have possessed. While it is certainly true that such things as a healthy body or birth into a prosperous life do not come to us on merit, it is a further stretch to say that this is true of all gifts and that it should always warrant an obligation to accept and keep whatever is given. It is not obvious that a terrible gift should be preserved regardless, for we can be entitled to find fault with the giver's choice of gift in spite of their intentions, and the grounds for this obligation are void if an undeserved good comes from no one in particular. We might believe that it creates obligations to help those with undeserved ills, but there is no inherent injunction against enhancing goods or indeed choosing preferable outcomes to what might otherwise have been given when there is no giver to speak of.

Separating gifts from the merely given would provide some philosophical justification for the gift motif we see mobilised in lay responses and objections to selective reproduction, where the key entailment is understood to be acceptance. As Scully et al. explain: "The notion of gift implies a lack of control over what is received. To speak about children as a gift is therefore to say that they should be accepted as they are, and that it is not appropriate to refuse them or to want to change their characteristics" (2006, 754). The emphasis is very much upon the virtues of parents as the recipients of gifts in their children, afforded them by divine or natural providence; as beneficiaries of the unmerited good of having children, parents should gratefully accept and welcome the gift of their progeny in whatever condition they are conceived. Yet these virtues have been a corollary of what, until recently, has been possible in reproduction; before ARTs were available, and under erstwhile religious presumptions of life's sanctity, the reality of procreation did not afford the possibility of selection, or indeed any of the virtues that we consider may now be attendant with it. The virtue was in *receiving* one's given child beneficently, which is to say without condition, since it was simply not within the gift of parents to select or stipulate the condition of the child. Sandel and Hauskeller endeavour to carry those same virtues over into an age in which it *is* now within the power of prospective parents to make such selections, in what looks like vain resistance to a new normative framework for reproduction that follows from the technologies that allow for ever greater procreative choice. Yet the language and code of the gift have not been left behind, but rather appropriated into a discourse of beneficence and obligation where parents are recast as givers rather than recipients of gifts in reproduction. In the following pages, I will show how two contemporary theorists of procreation do just that, one by way of opposing the creation of future generations altogether, and the other by positing an obligation to give life to future generations as well as possible.

Good Enough Lives

We have already briefly touched upon the difficulty of reconciling gift-edness with those lives that are unbearably brief and excruciating; it's hard to understand such lives as gifted because it is hard to understand such lives as good. While there are those who might faithfully maintain that there is good in such lives to which mere human beings are not privy, such conjecture is unintelligible at the bedsides of dying infants. If there is indeed a wider good or purpose according to some theology for which these lives are given, many people would flatly reject a giver with such designs in any case. As for the idea that parents receive a gift in the person of their child, regardless of that child's condition, the thought that some purposive intent lies behind the lives of desperately sick neonates is similarly objectionable. It seems clear that there are some lives so terrible that the attribution of goodness to their existence is plainly inappropriate, or at least relies on a notion of goodness unrecognisable to the vast majority of people. To paraphrase Hauskeller's qualification, and disregarding the idea that God has gifted such lives, no one would be able to say that they really are good, or could appear good to anyone. These tragic lives might be thought the exception to the intuition that to exist at all, in a condition of reasonable or at least bearable subsistence, is a good thing. However, there are those who deny that this is the case; that far from life being a gift, it is *never* a good thing to have been brought into existence.

This view proceeds from a curious feature of acts of procreation, which bring into existence individuals who would not otherwise have been. If, indeed, we are to talk of the giftedness of life, we need to acknowledge that such a gift is of a unique and different order to the kinds of exchanges and offerings that Hauskeller tests our intuitions against. This problematises our thinking about selective reproduction since the alternative to selection is not existence in some state other than that in which one could have been selected, but non-existence altogether. Similarly, to be selected—to be born at all—is to be brought into existence in the singular circumstance in which it were possible for one to exist. This has been called the paradox of future individuals (Kavka 1982), or non-identity problem (Parfit 1984). In this spirit, we can all reflect upon the precariousness of our own existence, the chance gene-alogy of our parents' unlikely passage to one another, not to mention their parents before them; the impulses, decisions and happenstance that led two individuals to meet, perhaps to fall in love, and by acci-dent or design to conceive a life at precisely the moment that they did. Any seemingly trivial deviation in the antecedent course of events would have spelled non-existence for us, just as any broader social or political variation effected by myriad strangers upon the lives of our forebears would have made for quite different people. Perhaps we might think there is somewhat less chance involved in the creation of children from arranged marriages, but even the juncture at which conception occurs is critical in the existence of one child rather than another. Given that one's unique genetic origins are a necessary condition of one's existence, if one had not been conceived at the time, in the circumstances and to the parents that one was, the only alternative state of affairs is that one would not have existed at all.

It is for this reason that we approach a difficulty in making claims about what is good or bad for the prospective individuals augured by the conscious and unconscious selections around reproduction. For if the alternative to existence in whatever state it is afforded us is non-existence, it is counter-intuitive to reckon that it would have been pref-erable never to have existed at all, except perhaps in the most extreme circumstances. It appears good to most of us that we were born, given that the alternative was not a differently embodied life in some other state or time, but no life whatever. Thus, claims that procreative actions or decisions can be bad for whoever is brought about in procreation have a problem: as John Robertson explains, "bringing unavoidably handicapped offspring into the world does not harm them because there is no way for them to be born healthy" (1994, 152). None of us can claim that we could or ought to have been born otherwise, for another birth would have resulted in quite another person. We might say that it would have been a better world, or one of less aggregate harm, had we never existed, but we cannot mean that it would have been better *for us*. Insofar as we suppose that the goodness or other-wise of an act can be evaluated as it affects persons, the act of bring-ing a person into existence—whom we can categorically say would not otherwise have been—can only ever be good for that person against the alternative of never existing at all.

However, although living may appear good to us all told, there is an asymmetry here that means it can be said that *not* living is not bad for us either. For in all the contingent moments approaching and indeed during our conception, it cannot be said that variations in events that would have resulted in the different possible world of our non-existence affected us for better or worse since we were simply not yet persons to be affected. Had my mother not been reluctantly persuaded to go to the party where she met my father, it would not have been bad for my siblings or myself since we didn't—and thereafter wouldn't—exist. There is no harm done. It may appear to me in retrospect that it was in my interests that she was persuaded, inasmuch as my life appears a good thing to me, but neither would it have been a bad thing for me never to have come into existence. The argument is complicated somewhat in assisted reproduction by the generation of multiple embryos, where those embryos which do not merit selection may yet be considered by some to have interests insofar as they have begun to exist *materially* on the threshold of the only lives they could have. However, assuming that the interests of embryos themselves, if they are coherent at all, do not carry the same weight as those of existing persons, we may still reject the notion that they can be harmed by a failure to implant them for gestation and birth. That is, even if we allowed that overlooking them for implantation could be said to be *bad for* the embryos, this is no more morally significant than saying a sperm cell's failure to fertilise an egg is *bad for* that sperm cell; there is no person-affecting justification for the expression of moral regret for whomever that cell might have created.

There are, as we have acknowledged, those most awful genetic conditions that do elicit our regret, where even the curiosity of consciousness is not thought to compensate for the horror of existence (see, for example, Archard 2004). But in such exceptional cases, where we compassionately consider it would be better that individuals not be brought into existence, the same problem of non-identity might seem to apply if an embryo is similarly indifferent to what is better or worse for it. Although it is not bad for the embryo not to be implanted that a resultant human being be brought into existence, can we say implantation *is* bad for the embryo—and worse for the potential emergent child to be born—given that the alternative non-existence is not a state that can be compared with anything? The most satisfactory answer for all practical purposes seems to be that of Joel Feinberg (1992), who interprets the assertion that one would have been better off not to have come into existence as a claim that "the preference for the one state of affairs over the other is a rational preference (…) In the most extreme cases … I think it is rational to prefer not to have come into existence at all, and while I cannot prove this judgement, I am confident that most people will agree that it is at least plausible" (17). According to this interpretation, we might expect only a small minority of lives to be supposed so bad as to be harmful in themselves; lives that could not plausibly be portrayed as gifts if we suppose that gifts must at least appear good to us. So much for the exceptions, but there is also an argument that the harmfulness of coming into existence is actually the rule.

According to Feinberg, if we can plausibly say that "non-existence in a given case would have been objectively preferable to existence … then any wrongful act or omission that caused (permitted) the child to be born can be judged to have harmed the child" (17). Taking a view on acts of selection then, and the technologies that allow us to predict and anticipate the medical condition of resultant children, we may say that it is morally blameworthy to permit the birth of a child in a condition over which non-existence is 'objectively preferable.' Feinberg's recourse to an idea of what is objectively preferable appears to be a necessary concession to a notion of the good that is independent of mere person-affecting criteria; after all, what is *subjectively* preferable is preferable to whoever is *subject* to preferences, and those who have not yet been brought into (morally considerable) existence are not yet subjects proper. But if we are to defer to a notion of what is objectively preferable at moments of reproductive choice, the burden of proof must surely fall upon those who have determined to bring a subject into existence at all. For, as David Benatar (2006) has argued, existence is always attended by significant harms, whereas nobody is subject to non-existence; thus not bringing a person into existence is not a deprivation to anyone of the pleasures of existence, but to bring someone into being is certainly to subject them to the assured and manifold pains of life. Although we might reasonably predict substantial and significant pleasures for the prospective lifetime of a given child, the absence of those pleasures in the alternative possible world of their non-existence is not bad since there is nobody for whom that absence is a deprivation. What is objectively preferable must be indifferent to the potential pleasures experienced by subjects not yet born, even if they were to live and consider for themselves that their lives really are good, for not bringing them into existence yields the certain good of avoiding the pain they would also no doubt have suffered. Given that the pains of existence are assured, and there is no harm in averting another's existence altogether, the most objective preference seems to speak against coming into life at all.

One need not be of an especially morbid disposition to concede that life is often utterly cruel and sometimes insufferable. Many people conclude, on the strength of the perceived horror of the world and/or their lives in it, that to no longer exist would be no deprivation in any case; that to give up existence is preferable to continuing to endure it. Many more consider the thought. In support of his anti-natal argument, Benatar describes a world of obscene suffering and travesty, of natural disaster, rapacious and unrelenting disease and all the dreadful terror that human beings inflict upon each other and the world around them. It is hard not to be persuaded that life, for most human beings who have yet existed, is indeed nasty, brutish and short.[2] Perhaps this makes all the more remarkable the widespread human instinct to cling to life at almost all costs, to regard existence even in the most miserable and tortured of circumstances as preferable to quitting it altogether. For Benatar, on whatever view one takes about what makes for quality of life, there is an invariable distinction between (a) how good a person's life actually is, and (b) how good it is thought to be. Recall that Hauskeller, though in a different context, mentions a similar distinction between what 'really is good' and 'what at least appears good to me.' According to Benatar, what merely appears good to me is no standard for what actually is good. He goes on to claim that we are predisposed to make favourable assessments of the quality of our lives by certain features of human psychology, and it is "these psychological phenomena rather than the actual quality of a life that explain (the extent of) the positive assessment" (64). This casts into doubt the reality and reliability of what appears good to me, since I am minded to find the good (and the gifted) in blithe disregard to my life's actual quality. For Benatar, that my life appears good to me is not sufficient to make a claim that it is a good in itself; in fact, it really is not good, nor were any prospective pleasures it might have held for me sufficient to invite the harms of bringing me into existence when the avoidance of those harms would have been good and the avoidance of me would not have been bad.

So far from being a gift, and even if we think our lives a good, being brought into existence is really a matter of dreadful bad luck according to the anti-natalist. Of course, some are certainly more unlucky than others, and this perhaps ought to motivate us to aid those condemned by accident of birth to lives more terrible than our own, but there should actually be a strong ethical presumption against bringing *any* lives about if we want to avoid causing harm. Benatar's central claim about the asymmetry of pains and pleasures in reproductive ethics, such that we can say the avoidance of pain is good while the avoidance of pleasure is not bad, leads him to advocate, apparently quite seriously, a phased and humane extinction for human beings. His dim view of the quality of lives speaks against the idea that we should be giving birth at all, at least if we intend the good in what is given. Benatar's commitment to what really is good or objectively preferable leads him to a conclusion where the disparity with what only *appears* good to human beings becomes superfluous on account of human beings being no more. That is, the question of the good is independent of human appreciation, and we should honour and seek the best state of affairs even if it does away with human beings altogether. The intuitive absurdity or pessimism of such a conclusion will be sufficient for some to oppose the premise of Benatar's argument. After all, if there are no persons for whom the distinction matters come our extinction, what really is good would be no more important than what appears good to persons; if the distinction is important, the subjects for whom the distinction matters ought to be preserved. And if the distinction is not *so* important, why not be content to leave human beings to the glad, if perhaps misguided, appreciation of lives that do appear good to them? But perhaps a rejection of the asymmetry underlying Benatar's anti-natalism must rely on an optimism that is beyond the scope of utilitarian reasoning alone.

No less a utilitarian than Peter Singer, in a *New York Times* discussion of Benatar's argument, resorts to simple optimism in resisting his conclusion: "In my judgement, for most people, life is worth living. Even if that is not the case, I am enough of an optimist to believe that, should humans survive for another century or two, we will learn from our past mistakes and bring about a world in which there is far less suffering than there is now" (2010). Singer seems unable to rebut Benatar's case on its own terms, repairing instead to his own hopefulness. Indeed, even if a world of far less suffering *were* brought about in another century or two, one wonders if the benefits for the inhabitants of such a world could be justification enough for the suffering of the subjects of the intervening years. It is quite some moral ledger to preside over, but it is in this optimistic spirit of seeking to bring about what is objectively preferable *for* human beings that some bioethicists advocate selection *of* human beings whose lives are objectively preferable.

Procreative Beneficence: Giving Best

Julian Savulescu does just this in proposing a principle of Procreative Beneficence, that where selection is possible prospective parents "have a significant moral reason to select the child, of the possible children they could have, whose life can be expected, in light of the relevant available information, to go best or at least not worse than any of the others" (2009, 274). Since the unselected cannot be harmed or wronged by not coming into existence, and children can be more or less disadvantaged in the possession of different qualities, it seems right to select children who will possess those qualities that are most advantageous in life and opposed by the fewest obstacles. It likewise appears morally suspect to select children whom we know will possess qualities that disadvantage them when, although *they* could not have been brought into existence otherwise, nor could they have been harmed by *not* being brought into existence in favour of creating another more advantaged child. Furthermore, it must also be morally suspect to refuse selection if it is possible, particularly where there is a known risk of passing on significant disadvantages, leaving the child's quality of life to chance.

Like Benatar, Savulescu is no stranger to controversy, for his views tend to challenge established parental norms and intuitions, though perhaps not quite so dramatically. While for Benatar no life is really good, or so good as to justify giving it, for Savulescu we must recognise that lives are more or less good and strive to bring about lives that are better than other possible lives. Tellingly, Savulescu pits his principle directly against his account of Sandel's argument, that "a child is a gift, to be cherished and loved for what she is. To be a good parent is to be prepared to accept and nurture one's child, regardless of that child's talents or disabilities" (274). Rather, where selection becomes possible, according to Savulescu, a good parent should not be indifferent to the qualities a child will inherit but ought to ensure that they possess qualities that allow their lives to go well, or not worse than the other possible lives at hand. So long as the means of selection is not morally problematic, prospective parents have good reason to seek to give birth to children with the best chance of flourishing. That is, they have more than simply freedom to do so with the technologies that are available, but ought also to feel a weight of obligation to use those technologies to create a child with preferable chances in life. The burden of justification is shifting from those who would seek to choose their child to weigh upon those who would not, so that we may hold as morally blameworthy those prospective parents who resist forms of selection that would vouchsafe the birth of the best possible child.

Savulescu finds no relevance for the language of giftedness employed by people like Sandel to object to enhancement and the exercise of parental discretion in reproduction. There is no external theological or cosmic intent for us to honour in conception, natural or otherwise, so why not make good, or best, of the children we can bring into existence? For Benatar, no life could be considered really good, or good enough to warrant inviting the manifold harms of existence; far from valorising the undeserved goods of existence, we should be more interested in averting the unmerited harms occasioned by birth. But in fact it seems that a notion of giftedness is still in evidence in both approaches. Peter Singer's comment on enhancement spells this out: "If there is no God, life can only be a gift from one's parents. And if that is the case, wouldn't we all prefer parents who try to make the gift as good as possible, rather than leaving everything to chance?" (2009, 279). This is basically an abridged version of Savulescu's argument, but tellingly Singer recasts the parents as the givers in the absence of a deity; if they can choose a more advantageous life than what is given left to chance, they are more virtuous givers for doing so. Although Savulescu dispenses with the language of gifts along with Sandel's case against selection, Singer helps us see how he retains the notion implicitly in the shift from considering the obligations of parents receiving gifts, to those of parents giving the gift of life to their children. In much the same way, David Benatar speaks against procreation altogether precisely because he argues that life is never a good thing to have been given. Though parents might try to make the gift as good as possible, it can never be really good, or good enough, to warrant giving it. For Savulescu, who does not question the objective preferability of existence, what matters is the intent of the givers in selecting the qualities of their prospective children. For Benatar, for whom it is always better never to have been, the quality of any given life is never sufficient to justify giving it; the greatest beneficence is in not giving at all.

Although there is rather more history to the anti-natal pessimism of Benatar,[3] it is important to recognise and reiterate the significance of the very recent developments that have given rise to Savulescu's notion of Procreative Beneficence. The moral understanding we find Sandel operating with is the product of a bygone reality where it was an impossibility for parents to entertain the thought of selecting the qualities of their offspring prior to birth. The virtues of gratitude, humility and acceptance were therefore quite proper in response to the hitherto unmanageable contingencies of procreation; parents understood themselves as receiving, more or less well, whatever gifts nature or God threw at them. However, where the various technologies of selective reproduction have allowed prospective parents to exercise some control in procreation, the virtues or obligations of giving life well now enter our thinking. The virtues are, amongst other things, a product of what is possible; where I cannot determine the outcome, I can only, in good character, accept any misfortune serenely. Likewise, to receive well a child out of the traditional contingencies of natural reproduction has tended towards parental ideals of unconditional love and hospitality. Yet

those contingencies have receded rapidly, so that the character of procreation is less a submission to fortune and increasingly a purposive gesture to the future, which demands more of prospective parents as givers.

On Savulescu's principle of Procreative Beneficence, parents ought
to embrace the opportunity to give life that confers the greatest benefit
upon its recipient. The word beneficence itself belongs to the language
of giving, of bestowing favour and generosity, translating from the Latin
bene facere—to do good. Such beneficence is entertained quite uncon-
troversially by prospective parents who consider carefully the juncture
at which they attempt to conceive, mindful of the disparate upbring-
ings and material resources they might be able to offer a different child
at different times. Savulescu is simply pressing for an extension of this
beneficence to the genetic inheritance that parents are also now able to
give their children by virtue of ARTs. If we consider it good to give a
child an environment in which it is expected their lives will go best, or
certainly not worse than others, it should also be equally commendable
to give a child a particular body or genome out of the same beneficent
purposes—and considered remiss to leave it to chance where selection is
possible. As Singer seems to imply in his comments, there is an ethic, or
at the very least an etiquette, to gift giving that means it is morally preferable, or more virtuous, to make the gift as good as possible.

Savulescu's principle of Procreative Beneficence marks the point at
which the liberalism of the new eugenics tips into more prescriptive
direction on reproductive choice, albeit awkwardly constrained by the
abiding commitment to individual liberty. To be sure, that liberty is
so fundamental as to be presently considered worth preserving against
the prospect of direct reproductive duress, even though it may result
in more disadvantaged and painful lives than might otherwise have
been; however, we can see in the clinical performance of pregnancy and
choice regimes of selective reproduction a certain coercion towards the
creation or selection of children whose lives can be expected to go better than others. The virtues invoked are pitted against the gift mentality
of many lay people, along with that of Sandel and Hauskeller, which is
rendered as an outdated moral deference to God or fortune, and a construction of reproductive parenthood as the receiving of unpredictable
and unwarranted gifts from a divinely ordered or natural providence.

In a sense, the critique of this position also amounts to an indictment of committing the naturalistic fallacy (Moore 1959), where the charge would be that the erstwhile contingent reality of reproduction has been mistaken for a normative edict on the nature of parenthood. Just because having children *was* a receiving of the unbidden, it does not follow that parenthood *ought* therefore to be an unconditional and passive acceptance of whatever child is conceived. The notion of giftedness tends only to be mobilised in reproductive bioethics to support this fallacy and the normative construction of parenthood as exemplified by the virtues of receiving gifts well.

The moral obligation to act beneficently in considering reproduction, either by avoiding it altogether according to Benatar, or selecting for lives of the greatest foreseeable advantage per Savulescu, takes a view of reproductive parenthood as more properly exemplifying the virtues of a giver. Although the language of giftedness is avoided, except in directly opposing it, there is, in Savulescu particularly, an advocacy and discourse of giving life well where the agency facilitated by ARTs allows parents to exercise and evidence good intent in the qualities they give to their child. The gift is present both in Sandel's objections to selective reproduction and Savulescu's endorsement of it; the difference is that Sandel figures parents as properly receiving gifts in their children, and Savulescu sees parents as newly capable of giving them, and being morally superior in seeking to 'make the gift as good as possible.'

The ability to select or engineer so that the goodness of one's procreative will can be made incarnate in the characteristics of one's children seems to secure the very capacity for beneficence in the narrative of optimism that accompanies ARTs.[4] Newly cognisant and able to choose the child to be given birth, and to make that gift as good as possible, we can now find that the parental refusal of such choices is regarded as a morally blameworthy rejection of the possibility of being beneficent. The parental ethic of unconditional welcome becomes a well-meaning but anachronistic category error of thinking procreation as accepting gifts, when the purposive selections offered by ARTs now make it more properly the chance to give them. Beneficence is gauged against the practices of what is possible, so that to decline the possibility of giving life better, or a better life, is to eschew a certain moral responsibility to future generations. The story about procreation that emerges is that where once it was an exposure to the vagaries of fortune, and thereby culturally or religiously interpreted as a gift from God or nature, now that it is subject to the wilful intention and design of creators we can exercise beneficence and discern the virtues of gift-givers in choosing children.

In the following chapter, I will show how the advocates of genetic enhancement and selection depict unassisted or natural reproduction as a creation lottery, against which ARTs allow the prospect and obligation of securing less harmful outcomes than chance would afford; the colonisation of the natural by the just (Buchanan et al. 2000), where the mother is associated with nature and the medical technologies of intervention augur justice. This portrayal, I will argue, colludes with a long and ignominious patriarchal history of effacing and denigrating the maternal body, and forgetting the gift of life delivered only by mothers. Reviving the thought that life has always been a gift, not received from God or out of a morally indifferent vacuum, but given by mothers, we can regard the significance of selective reproduction as not giving life anew with a novel access to beneficence furnished by technology, but rather giving life differently. A keener regard for this difference, informed by a more critical appreciation for the matricidal thinking that is still reproduced in reproductive bioethics today, allows for a perspective that articulates what may be at stake in the shift from giving life to strangers, to giving life to chosen future kin.

Notes

1. For a thorough account and analysis of moral luck in reproduction and its implications for principles of justice, see Yvonne Denier (2010) "From Brute Luck to Option Luck? On Genetics, Justice, and Moral Responsibility in Reproduction." *Journal of Medicine and Philosophy* 35(2): 101–129 (Denier 2010).
2. Though levels of violence have been in decline over millennia, so that existence in the present age is reckoned on some accounts a far preferable prospect to living in other historical periods—see Steven Pinker (2011) *The Better Angels of our Nature: The Decline of Violence in History and its Causes*. London: Allen Lane (Pinker 2011).

3. For example, Sophocles, E. Grennan and R. Kitzinger (2005) *Oedipus at Colonus*. Oxford: Oxford University Press. Also, Schopenhauer, A. (1900) *Studies in Pessimism: A Series of Essays*. [S.l.], Swan Sonnenschein (Schopenhauer 1900; Sophocles et al. 2005).

4. For an account of the clinical and professional narratives of genetic technologies, medical screening and disability, see Tom Shakespeare (1999) "Losing the plot? Medical and activist discourses of contemporary genetics and disability." *Sociology of Health & Illness* 21(5): 669–688 (Shakespeare 1999).

References

Archard, D. "Wrongful Life." *Philosophy* 79, no. 309 (Jul 2004): 403–420.

Benatar, David. *Better Never to Have Been: The Harm of Coming into Existence*. Oxford: Clarendon, 2006.

Buchanan, Allen E., Dan W. Brock, Norman Daniels, and Daniel Wikler. *From Chance to Choice: Genetics and Justice*. Cambridge: Cambridge University Press, 2000.

Caplan, Arthur. "Good, Better or Best?" In *Human Enhancement*, edited by Julian Savulescu and Nick Bostrom, 199–209. Oxford: Oxford University Press, 2009.

Denier, Yvonne. "From Brute Luck to Option Luck? On Genetics, Justice, and Moral Responsibility in Reproduction." *Journal of Medicine and Philosophy* 35, no. 2 (2010): 101–29.

Feinberg, Joel. *Freedom and Fulfillment: Philosophical Essays*. Princeton, NJ: Princeton University Press, 1992.

Harris, J. "Sex Selection and Regulated Hatred." *Journal of Medical Ethics* 31, no. 5 (May 2005): 291–94.

Hauskeller, M. "Human Enhancement and the Giftedness of Life." *Philosophical Papers* 40, no. 1 (2011): 55–79.

HFEA. "Sex Selection: Options for Regulation. A Report on the HFEA's 2002–2003 Review of Sex Selection Including Discussion of Legislative and Regulatory Options." http://www.hfea.gov.uk/docs/Final_sex_selection_main_report.pdf, 2003.

Hyde, Lewis. *The Gift: How the Creative Spirit Transforms the World*. Edinburgh: Canongate, 2007.

Kass, Leon. "Ageless Bodies, Happy Souls." *The New Atlantis* 1 (2003): 9–28.

Kavka, G. S. "The Paradox of Future Individuals." *Philosophy & Public Affairs* 11, no. 2 (1982): 93–112.

Moore, G. E. *Principia Ethica*. Cambridge: The University Press, 1959.

Parfit, Derek. *Reasons and Persons*. Oxford, Oxfordshire: Clarendon Press, 1984.

Pinker, Steven. *The Better Angels of Our Nature: The Decline of Violence in History and Its Causes*. London: Allen Lane, 2011.

Robertson, John A. *Children of Choice: Freedom and the New Reproductive Technologies*. Princeton: Princeton University Press, 1994.

Sandel, Michael J. *The Case Against Perfection: Ethics in the Age of Genetic Engineering*. Cambridge, Mass; London: Belknap Press, 2007.

Savulescu, J., and G. Kahane. "The Moral Obligation to Create Children with the Best Chance of the Best Life." *Bioethics* 23, no. 5 (Jun 2009): 274–90.

Schopenhauer, Arthur. *Studies in Pessimism: A Series of Essays*. [S.l.]: Swan Sonnenschein, 1900.

Scully, J. L., T. Shakespeare, and S. Banks. "Gift Not Commodity? Lay People Deliberating Social Sex Selection." *Sociology of Health & Illness* 28, no. 6 (Sep 2006): 749–67.

Shakespeare, Tom. "'Losing the Plot'? Medical and Activist Discourses of Contemporary Genetics and Disability." *Sociology of Health & Illness* 21, no. 5 (1999): 669–88.

Singer, Peter. "Parental Choice and Human Improvement." In *Human Enhancement*, edited by Julian Savulescu and Nick Bostrom, 277–291. Oxford: Oxford University Press, 2009.

———. "Should This Be the Last Generation?" *New York Times*, June 6, 2010.

Sophocles, Eamon Grennan, and Rachel Kitzinger. *Oedipus at Colonus* [Translated from the Ancient Greek.]. Oxford: Oxford University Press, 2005.

3

Creation Lottery and Mother Trouble

Abstract A close reading of a footnote in a paper by bioethicists John Harris and Julian Savulescu points to their complicity with a history of deprecating the gifts of the maternal body, with their depiction of the risky and profligate 'goddess' creation lottery. This chapter goes on to draw out their implicit inheritance of a Platonic and theological aversion to acknowledging the gift of life from mothers, and appropriation of birthing language and metaphors in creating sources of metaphysical or spiritual meaning. The uncritical inheritance of this patriarchal tradition informs a bioethical account of procreative beneficence as a novel capacity to give life well against the amoral, blameworthy bodies of women.

Keywords Birth · Lottery · Creation · Maternity · Goddess · Feminism

In 2004, two prominent bioethicists John Harris and Julian Savulescu co-authored a paper entitled 'The Creation Lottery: Final lessons from natural reproduction: Why those who accept natural reproduction should accept cloning and other Frankenstein reproductive technologies.' The paper attempts to establish what a commitment to the permissibility of natural reproduction entails, particularly given the fact

© The Author(s) 2017

S. Reader, *The Ethics of Choosing Children*, Palgrave Studies in Ethics and Public Policy, DOI 10.1007/978-3-319-59864-2_3

55

that natural reproduction involves a very high rate of embryo loss so that creation *in vivo* as a result of sexual intercourse might be thought of as a lottery. As they explain:

> Natural reproduction is a practice that involves the creation of a population of embryos for the purposes of creating a new human being and that involves the unavoidable death of 80% of those embryos. To put this more generally, a creation lottery involves the creation of a population of embryos for the purpose of creating a new human being, and this practice involves the unavoidable death of some of these embryos. (2004, 90)

Those who countenance natural reproduction, they argue, are thereby also logically committed to running other relevantly similar creation lotteries where some embryos perish in the course of creating a new human being—and where 'unavoidable' is defined as "given the current state of the world including knowledge, values and intentions of human agents" (91). The authors go on to contend, therefore, that early forms of assisted reproduction, PGD, the voluntary assignment of embryos for research and reproductive cloning must all in principle be permissible for those who endorse the morality of the creation lottery in natural reproduction.

It is an unsurprising paper, save for a rather extraordinary and gratuitous footnote at the end, where Harris and Savulescu declare: "After a difficult and fractious courtship, we have both learned to love or at least admire the perfection of one goddess—the creation lottery. Those who believe that natural reproduction is permissible are committed to the worship of this goddess" (95). The footnote itself appears at the close of the paper where Harris and Savulescu briefly conclude with a discussion of 'method' in bioethics, bemoaning a general "lack of constructive dialogue" in the discipline, and presenting their own work as the "constructive rational discourse" that they uphold as exemplary. This has been criticised elsewhere (Holm 2004), but suffice it to say that, in this paper at least, the "embryo rightists" they target are little more than straw men; after all, they had to coin a name for their own opponents and therefore do not in fact engage in constructive dialogue with the real people whom they oppose. The footnote appears only as an unnecessary restatement of what we know their position to be, except that it is expressed in such surprising hyperbole, and incongruous with the "rational discourse" they have been advocating.

Of course, we are not to suppose for one moment that Harris and Savulescu are sincerely invoking, much less exalting, a deity as the source of life by characterising their creation lottery as a goddess. Rather, it seems likely that the footnote is intended, in part, to make fun of those who valorise nature or worship some supernatural power behind the creation of life. Harris and Savulescu state very clearly in their final comments: "Bioethics is not about conversion (that is the province of religion) or convincing others that one is right. It is about discovery of the truth and gaining knowledge" (95). Their rebuke to faith traditions emphatically disowns religious thought of any claim to rational discourse, undercutting the relevance of any religious insight into the objects of bioethical inquiry.[1] The scientific account of natural reproduction as a chance and embryo-spoiling process contradicts the reverence with which many religions regard it, as the divinely gifted and therefore only morally legitimate form of reproduction. Thus the reference to their creation lottery as a goddess appears ironic at best, and completely derisive at worst. According to Harris and Savulescu's faux-religious footnote, a belief that natural reproduction is permissible should actually commit people to a profligate feminine deity that is a far cry from that of the so-called 'embryo rightists.' They perversely apotheosise this creation lottery in order to expose the absurdity of venerating the brute speculation of nature.

Yet just as revealing is the way that the footnote frames their relationship to gaining knowledge through bioethical method. They describe truth somewhat Platonically as something to be discovered, as though it were an alienable thing out there, vulnerable only to the bioethicist disinterested and yet intrepid enough to find it. What is more, they present that knowledge—in this case, that of a creation lottery—as the *feminine* object of their joint intellectual advances. They describe their method as "courtship" where they are the active male suitors and the knowledge they gain a goddess whom they may love, or at least admire if they cannot quite deign to endearment. Perhaps love would imply a loss of control, a clouding of judgement, a commitment to one truth that might compromise their autonomy to court others. In any case,

this account of truth—albeit facetious—is notable for being both passive, female and the object of a "fractious" romancing by two men. The
unruly, irascible truth playing hard-to-get! It is surely a curious way
of figuring the approach to gaining knowledge, and appears to rely
on quite dated and obviously patriarchal assumptions about gender
roles. Certainly, if truth is a goddess to be courted, bringing to mind
courtship's overtures to traditional marriage, then the heterosexual
man is surely in a privileged position—a familiar story in the history
of Western thought. Women and non-heterosexual men will find that
truth cannot be the object of *their* affections; philosophy and bioethics
are not for them.

Of course, it is important to remember that these remarks appear
only in what is a rather offbeat footnote, but the close reading of such
marginalia can help to draw out elements of a text that are revealing
of underlying assumptions and attitudes not explicitly stated in the
main body of the work.[2] In any case, Harris and Savulescu also make
some factual assertions in the body of their paper that also imply a
particularly casual view of sexual difference. In staking out the "potentially relevant moral features of natural reproduction" they declare with
audacious ease that "natural reproduction is voluntary" (90), happy to
state this as a matter of abstract fact without further qualification. The
question of how often and *for whom* it is voluntary—and what voluntary might mean for men and for women—is apparently moot. If they
consider any concession to forced intercourse or involuntary conception through rape, for instance, as potentially morally relevant, they
fail to acknowledge it. The premise that women—and, for that matter,
men—always volunteer for the sex that occasions natural reproduction
is simplistic or deeply complacent. Similarly, the authors nonchalantly
contend that "there is an alternative to natural reproduction: childlessness through contraception or abstinence" (90), without considering
that choosing childlessness might mean different things for men and for
women given that the experience of gestating and bearing children is
particular to women. Of course, these features are so described in order
that Harris and Savulescu can argue that embryo loss in this particular
creation lottery is avoidable insofar as natural reproduction is avoidable.
But to assert that sex is straightforwardly voluntary, sexual difference

morally neutral (Harris 2007, 144), and childlessness an uncomplicated alternative to procreation, appears to radically overlook the sexed and different experiences of men and women in procreation.[3]

One other possibility may be that the so-called 'goddess' creation lottery courted by Harris and Savulescu is homage, of sorts, to the fact that human beings are of woman born—characterising their deity as female at least undermines the perversity of a traditional religious belief that life is the gift of a paternal God. However, the creation lottery they propose is patently void of any agency whatsoever, so at the point when life is wrested by philosophers from the gift of a male God, it is actually not to suggest or affirm our coming from a maternal body, but rather, instead, to reduce reproduction to a destructive and capricious game of chance. It is only when human beings can take charge over the accidents of existential fortune and misfortune that proceed from this lottery, for Savulescu, that we can begin to conceive of (or with) beneficence in reproduction at all. What is not acknowledged is that this rendering of natural reproduction as creation lottery is also an account of the maternal body—in which the lottery of conception is held—as inefficient, unsafe and wasteful. Anything but beneficent. A creation lottery deified as goddess is no compliment to the maternal body then, since that very lottery is what Harris and Savulescu contend we have an obligation to overcome by seeking to select "the best people—that is, people with the longest and best-quality lives" (2004, 93). Where technology promises the incarnation of reproductive design and aspiration, the bodies of women are rendered sites of compromise and jeopardy; of unsafe keeping. They *may* ally the beneficent purposes of prospective parents and clinicians, but void of the assistance of the technology that promises the expression of those purposes, women's bodies are mere bearers of precarious embryonic fortune. And though human beings are of woman always born, only now that technology permits the subjection of reproduction to the realisation of deliberate intent or selection do we begin to acknowledge beneficence—and we find maternal bodies ignored, or even pitted *against* it.

The practical danger of this discourse is that it becomes reflected in the practice of reproductive technologies and the experiences of those who encounter them. Many women who undergo the technologies of the tentative or assisted pregnancy attest to an estrangement from their femininity and a clinical disregard for their part in the process. "I felt a total lack of femininity. I reverted to a sort of neuter...I felt terribly spayed. It was quite loathsome...I did not really believe my body was there" (Birke et al. 1990, 63). Such experiences chime with the view expressed by some feminists (e.g. Corea 1985; Greer 1999), that reproductive technologies can often disempower women and distance them from the process of giving birth. Subject to the interventions and surveillance of the medical techniques that are employed to assist and regulate pregnancy and labour, women may actually feel a *loss* of control and alienation from their life-giving power, reduced to little more than incubators or walking wombs. For Adrienne Rich (1977), the exclusive power possessed by women to give and support new life has exercised a reactionary male desire to exert power *over* the reproductive process and control the bodies of women. This unwillingness or failure to recognise the power and gift of maternity becomes present in the medicalised practices and discourse of reproductive technology, and is propped up by a bioethics that reduces the unassisted maternal body to little more than a slot machine.

Likewise, there is a dissonance in the distribution of credit and culpability when assisted reproduction either succeeds or fails, for in success the achievement is constructed as belonging to medical science along with its practitioners, yet in failure the fault falls back upon the blameworthy body of the woman. Karen Throsby (2004) has researched the experiences of prospective parents undergoing IVF treatment, where women still tend to assume the primary responsibility for pregnancy in terms of seeking medical information and assistance, and policing their own bodies. Tellingly, she found the following:

> When IVF succeeds, this burden of labour and responsibility is obscured with success attributed to the technology (and its providers) – hence, the IVF baby. But when treatment fails, women are written back into the narratives as implicitly, or even explicitly, culpable for that failure, absolving the technology, its practitioners and also the male partners of that responsibility. (134)

So success is credited to medical power over reproduction, but failure pinned on the fickle faculties of the mother's body. It is to the technology and its practitioners that gratitude and indebtedness is directed, but in the practice and discourse of that technology it somehow manages to keep acclaim for the (limited) successes while avoiding blame for failure. This accords all too neatly with a long history of constructing maternal culpability for reproductive failure, the birth of children with disabilities or conception of too many girls (Stonehouse 1994, 35). Women are made accountable for the failure to give birth, or give birth well, but we are not encouraged to recognise any debt to women or maternal bodies for what they *do* give and have given, even in the most extraordinary and challenging of circumstances. As Throsby observes, this account "of female insufficiency as an explanatory framework for IVF and its failure is a very sobering finding" (159), and although reproduction may now be performed with novel technologies, some inherited attitudes clearly still prevail. These attitudes continue to obscure or overlook the beneficent agency of the maternal body, whilst at once holding it alone answerable for reproductive failure; a discrepancy that, among other things, seems to indicate a strong and deep resistance to acknowledging the generosity of maternity.[4]

Savulescu and Harris argue that if reproductive technologies "become more efficient or safer than natural reproduction (i.e., would have a lower embryo wastage or fewer deformities), then they would be preferable to natural reproduction" (2004, 93–94). There is no consideration of whether natural reproduction might yet be preferable for mothers, or admission that these preferences are morally relevant; efficiency and safety here are measured only by outcomes for embryos and potential children. The moral relevance of the maternal body is written out completely, indeed their invocation of Frankenstein to designate the various technologies of reproduction implicitly reinforces their account of its insignificance. In their paper they refer collectively to all the novel means of choosing children as Frankenstein technologies, describing the science of prenatal genetic selection and enhancement without a single reference to the maternal body that will ultimately bear the child. As we know, Frankenstein's creation was accomplished in the story, somewhat imprecisely, without being borne by a mother, and although Harris and

Savulescu may be writing with some glibness, the trope of Frankenstein encourages a further forgetting of the mother in our evaluation of reproductive technologies.

This forgetting is evidenced neatly—and remarkably—in certain media portrayals of ARTs, for instance a Daily Mail report in 2017 on a procedure called in vitro gametogenesis (IVG) by which it could be possible to create sperm and egg cells from skin tissue. That's to say, it could be possible to conceive a child in the laboratory from sex cells generated independently of functioning reproductive organs. The Mail's headline ran: "Babies made without mothers 'will come sooner than we think', leading scientists warn after study discovered how to make embryos from skin cells" (De Graaf 2017). The article's focus on "mother-less babies" and the possibility of conception using the cells of two men reduces procreation to the act of conception alone; in fact there is no mention whatever of the embryo's fate beyond conception and before birth. The inconvenient nine months or so of nascent life within, sustained and borne by a maternal body. She is entirely absent from this account, as though pregnancy were incidental, or that her participation as a surrogate were a given; either way, at the first prospect of contemplating conception without a woman we take the opportunity to completely write her out. The rest of this chapter will extract some notable examples of this forgetting or erasure of the maternal body from our inherited intellectual and cultural history, suggesting that Harris and Savulescu's apparently radical claims actually echo a very familiar story.

Plato: In Spite of Birth

Harris and Savulescu's goddess creation lottery seems to be a mimetic allusion to the cults of a Mother Goddess or historical theories of Mother Right or matrifocal society, such as those espoused by Johann Jakob Bachofen (1967), and Lewis Henry Morgan (1877). That is, the veneration or recognition of maternity where in fact there is nothing but chance to be worshipped in the lottery of reproduction. They could not help but permit themselves this ironic gesture, though it reminds us that the maternal body is written out of the story. One thinker of late who has looked to explicitly *write in* the philosophical significance of the universal human condition of birth from the maternal body is Adriana Cavarero. In her celebrated book *In Spite of Plato* (1995), she shows how the female subject who seeks out mythic figures of her own sex, encounters only stereotypes and crude male configurations of the feminine. She finds herself already imagined according to an order where woman is the object of the (masculine) other's thought rather than a subject in her own right; where women occupy roles in relation and reference to the figure of the universal male subject at the centre of creation. Where Bachofen found in the mythical tradition "immediate historical revelation" (73) of an ancient culture of Mother Right that was subsequently overturned by what he saw as a superior and civilising patriarchal order, Cavarero reads a distinct and insidious lack of female subjectivity that denies adequate representation of sexual difference. Though she affirms with Bachofen the shift from an archaic culture of matriarchy to the patriarchal symbolic order that we now inherit, her project is to investigate "the traces of the original act of erasure contained in the patriarchal order, the act upon which this order was first constructed and then continued to display itself" (5). She takes, therefore, the work of Plato, to which the entire European philosophical tradition has been famously referred as footnotes (Whitehead 1929, 39), as her literary context and the moment to revive the female subjectivity closeted there.

Cavarero takes some of the female figures referenced in Plato's texts and the surrounding Greek myths, and carefully works them free of the male imaginary in which they are contained. One such figure is that of Demeter, the Great Mother who possesses the secret of fertility and life, seemingly in the tradition of the Mother Goddess posited of original matriarchal religion. Yet, in Plato's account, "It appears that Demeter was named after the gift of food she gave (*didousa*) as a mother (*meter*)" (*Cratylus* 404b). For Cavarero, this reference is symptomatic of the effacement of Demeter's more sacred and more dangerous status as the source and repository of the secret of life, a secret imparted in nature at her discretion. According to the myth, Demeter's daughter Kore is abducted from her by deception and taken to the underworld, or kingdom of the dead, to be wedded to the god Hades. In her

grief and desperate search for her daughter, Demeter ceases to generate and the lands and living things of the earth fall sterile, posing the
threat of extinction to humankind and the world. Hades is persuaded
to return Kore to her mother for a period of time every year, a period
which returns warmth and fertility back to the earth for a season; but
when Hades reclaims her again each year the earth turns cold and sterile once more. The central theme of the myth, as Cavarero points out,
"is the power of the mother, which is inscribed in all of nature as the
power both to generate and not to generate. This is an absolute power
that presides over the place from which humans come into the world
and over nothingness, as birth-no-more" (1995, 59). In Plato's remark,
however, this meaning is lost. To be sure, the mother is credited with
providing food and nurturing care for life, but no longer acknowledged
is her absolute power to give or withhold life itself—a power, it is worth
recalling, that philosopher David Benatar suggests we ought to appropriate to eventuate our extinction.

Cavarero finds Plato's fingerprints elsewhere in our inherited disavowal of coming from a mother, not least in the seminal assertion of
other worlds or more certain realities beyond our immediate, embodied condition. We are familiar enough with the tale of Socrates' death,
which he welcomes, as the untethering of his soul from the prison
of the body. Cavarero explains: "Plato's thesis is this: by leading one's
thinking toward eternal objects suitable to pure thought (pure ideas),
philosophy *unties* the soul from the mortal body. Therefore those who
lament the fact of death, which is the definitive untying of the soul
from the body, are bad philosophers" (22–23). This sets the precedent
for what philosophy—or good philosophy—consists in; thought that
outstretches and exceeds the vicissitudes of the material world to contemplate metaphysical ideals and forms. An overcoming of the *merely*
physical to dwell upon and among the timeless truths set apart from
the transient and accidental matter of embodied experience. It is with
these eternal objects that the soul belongs, so that through philosophy it may abide again, albeit imperfectly, in its proper sphere until it
is released definitively from the "embarrassing dead weight" (27) of the
body. The separation and opposition is thus established between the
material (maternal) and the metaphysical, the biological and the divine,

body and soul (or mind); the soul experiences its sojourn in the body as an incarceration, and philosophy, or pure thought, is conducted in constant anticipation of escape through death. Physical birth then, is the commencement of an inconvenience or burden, arresting our communion with what is true or most real by anchoring the soul in the vulnerable, untrustworthy and dependent body. Philosophy, or the pretension to metaphysics, deprecates the life of the body and the significance of our earthly incarnation, orienting us instead toward the possibility of a more authentic accord with truth in our release from the material world in death.

Plato's most audacious move is to appropriate the language and imagery of birth from the physical world to commend precisely the metaphysical world he would have upstage it. What is more, he takes a woman for his mouthpiece in the person of Diotima in the *Symposium*, who discusses the meaning of love as a means of directing human contemplation toward the divine. Love is therefore separate from procreation, but the very spectre of procreation in heterosexuality makes it less devotional than (male) homosexual love; the prospect that potentially procreative heterosexual love may produce or be a means to something other than itself makes bodily fertility and maternal power in particular a liability. By contrast, homosexual love can be an end in itself, can find complete fulfillment in itself and allow the lovers to contemplate the other's beauty and the divinity of their love uncompromised by external ends. According to Diotima in the *Symposium* then, love is, in effect, "a giving birth in beauty, both in body and in soul" (206b), lending the language of the maternal power of which *she* is possessed to a philosophy that actually denies it to her. In Diotima's speech, as Cavarero explains, "maternal power is annihilated by offering its language and vocabulary to the power that will triumph over it, and will build its foundations on annihilation itself" (94). It is philosophy that breeds the most authentic, the most beautiful and immortal offspring: the ageless and incorruptible virtues and truths conceived in concert between men, like Harris and Savulescu, fertile in wisdom and justice. Human reproduction is not only an entrapment of the soul into flesh, but itself detracts from the communion of souls in physical love for its own sake. Maternal power as the means of carrying out sexed reproduction becomes a barrier to true fertility, while "the pregnant, birth-giving male, like the male who practices midwifery, stands as the emblematic figure of true philosophy" (Cavarero 1995, 92).[5]

The appropriation of birthing metaphors and language in Platonic philosophy is not in homage to the power of the maternal body then, but is rather in order to deprecate it, to play down its significance. Nowhere is this more evident than in Plato's allegory of the Cave, where he lets Socrates describe a group of people who live chained within a cave, captive to the shadows cast on the wall before them by models of people and animals passing in front of a fire at their backs; these prisoners, who can see neither themselves or each other, take the shadows to be the things themselves, and the echoes of the puppeteers' exclamations to be their voices. The forms the prisoners perceive on the wall before them—which is in fact but the shadow play of the puppets behind—they take as reality, and live thereby according to illusion. In Socrates' tale, one of the prisoners is freed from his chains to painfully confront the counterfeit forms and flames that he had taken for reality; then, forced out of the cave into the daylight, he is able to see the sun—the one and true source of the seasons and the years, and ultimate cause of all that he and his companions had been seeing. The myth serves as a picture of the philosopher's progress from ignorance to knowledge, a laboured emergence from the shackles of corporeal existence and corruptible perception to the true reality attainable by reason, detached from the given flesh of the body.

Plato's Cave allegory has been taken as formative, if not the foundation, of Western metaphysics and culture. What we take for reality by virtue of our senses is an illusion; rather, we arrive at truth by disowning the irrelevance of corporeal life and seeking through pure reason and contemplation the sources of genuine knowledge. Through philosophy, the soul emancipates itself from the physical bonds that fetter it; the hero philosopher emerges from the cave to possess the authentic truths that the body denies to him. Along with further derogation of the body and its imperfect knowledge, downplaying the maternal agency in our bodily inception, the tale of emergence into truth here offers plain echoes of nativity and parturition: "Plato's own extensive use of birth metaphors invites us to interpret the cave as a womb...Plato operated with two concepts of birth: the physical birth, whereby the soul is unified with a body, and the spiritual, which signifies the soul's emancipation from physical bonds. It is just such a spiritual birth that is described in the cave myth" (Songe-Møller 2002, 118). Just as he has Diotima endorse a love that excludes her, so Plato's cave allegory exploits the female body and its possibilities as a source of symbols and metaphors to explain a knowledge that diminishes that body; a spiritual birth that supersedes the mere birth of the flesh. Luce Irigaray's rich and detailed rereading of the cave myth draws out the way that "the womb has been played with, made metaphor and mockery of by men" (1985, 263). A mockery that we might observe is continued by Harris and Savulescu's goddess creation lottery. At the seminal moment and trope of Western philosophy we find the imperative to *deliver oneself* free of the limitations and flaws of the body with which one already lives—or to be delivered by the elite pederast-philosopher. It is as though one's soul were subjected to embodiment, and mothers, far from conferring some benefit with existence, were mere accessories to the nuisance of it.

Born Again with Christ and Anti-Christ

As Martin Heidegger explains, Platonic thought tore a chasm "between the merely apparent beings here below and the real Being somewhere up there. Christian doctrine then established itself in this chasm, while at the same time reinterpreting the Below as the created and the Above as the Creator" (2000, 111). Thus the imperative under both Platonism and the age of Christendom that followed was to concentrate upon the life of the mind or spirit, or submit to the authorities thereof, and decry the embodied state that keeps us at a distance from the true or most real. Friedrich Nietzsche famously described Christianity, amongst other things, as "Platonism for the people" (1973, 31), recognising in both traditions the fixing of a gaze upon the Good, or Spirit, away from the corporeal world and untrustworthy sensory realm. Nietzsche is often identified as bringing the curtain down on the Western metaphysical tradition of thought exemplified by Plato and aped by Christianity, so that in declaring the death of God he is also asserting the abolition of an independent or true world beyond the realm of appearances. Philosophers, moralists and priests alike, in affirming their recourse to a world beyond this world, are really saying no to life as it is: a complex, fluid and material web of interpretations.[6]

Rather like the Greek philosopher, the good Christian is historically instructed to abandon the world of their birth in order to establish their abode in pure faith; they must transcend mere physical emergence out of the womb and labour of their mothers with a birth of the spirit to become divine. A second birth which disinherits the mere maternal delivery of the flesh and is possible through God alone: "Except a man be born again he cannot see the Kingdom of God" (John 3:3). In the Christian tradition, as Kahn explains, "females become synonymous with first birth, or birth of the biological self, which signifies mortality and inferiority…The virgin birth becomes an early new reproductive technology in which the human mother becomes a vessel for the gestation of an immortal child who belongs not to her but to the father" (1995, 165). The Bible follows Greek myth in diminishing the maternal power of generation for bringing life into a lower and corrupted world that must be overcome and supplanted by a more important manner of creation. They both establish a distinction between birth from below— out of darkness, nature, emotion—and birth from above—out of light, culture, spirit or reason. "Maleness belongs to birth from above, which, culturally and historically, becomes a more prestigious form of creation than birth from below. Thus, in both Greek myth and the Bible, which are the roots of Western tradition, creation stories diminish the generative role of the female by dismissing birth from below" (Kahn 1995, 167). Christianity posits direct and divine intervention into birth from below to have Christ delivered *of* woman but at a remove from maternal generation and generations by sparing him the indignity of being created *by* woman. He promises that we too can outmode or succeed our birth from below to attain a place in the Kingdom of God if we are born again in Spirit.[7]

Nietzsche's philosophy, self-described as 'inverted Platonism' (Heidegger and Krell 1979, 154), seems to bring us back down to earth, auguring the possibility of finally acknowledging the significance of our embodied creation in maternity. He addresses the 'despisers of the body' in *Thus Spoke*

Zarathustra, and affirms against their dualism that "the awakened, the enlightened man says: I am body entirely, and nothing beside" (1961, 61). Nietzsche opposes those Platonists and priests who set us against the wisdom, intelligence and multiplicity of the body to commend and theologise the mind or soul, and a second birth or rebirth into a superior realm of understanding or being. It is, rather, with our bodies that we compare and create; our bodies that interpret the world in its pain and pleasure, giving rise to thought and reflection (see Diprose 2002). In one of his most famous passages, entitled 'Of the three metamorphoses,' Nietzsche writes of the transformation required of human beings to overcome Christian moral interpretations of life, moving from the weight-bearing spirit of a camel, to the lion that wills to be "lord in its own desert," and finally, and most redemptively, to become a child: "a new beginning, a sport, a self-propelling wheel, a first motion, a sacred Yes" (1961, 54–55). For Giles Fraser (2002), this final metamorphosis is actually parasitic on the spiritual circumstances that Nietzsche wills to overcome. It echoes, of course, the idea expressed in the New Testament: "unless you change and become like little children, you will never enter the kingdom of heaven" (Matthew 18:3). But by invoking the new beginning of childhood as the last transformation, Fraser argues that Nietzsche "takes the 'born again' metaphor more literally than was intended" (104). Recalling the biblical story, Jesus does not instruct Nicodemus to be born again as a child, to return to his mother's womb, since he asks us to face up to our past. In fact, the Greek word he uses has the double meaning of 'again' and 'from above' so that Nicodemus's query about returning to his mother's womb is actually a kind of joke. One cannot be born again physically, but neither does Jesus want Nicodemus to forget his past. As Fraser points out, "To be born again as a child is to be born again as an amnesiac" (105). Nietzsche tells us that the preying lion must become a child, but without recourse to the 'from above' with which he has dispensed, this transformation of the spirit in forgetfulness just looks like the impossible physical birth 'again' as Nicodemus (mis)understood it.

Fraser's contention is that, despite his ferocious attacks on Christianity, Nietzsche himself was preoccupied with the question of human salvation. He borrowed from the Christian past that he rejected in a flawed but "complex and sophisticated process of theological sampling" (2), in pursuit of redemption for human beings, albeit in a world without God. Nietzsche is stuck, rather like the lion, preying on the freight of the weight-bearing spirits he savages. The requirement that the metamorphosis should go beyond the lion to the 'new beginning' in the child betrays Nietzsche's instinct for some redemptive consummation of spirit that, Fraser argues, is ultimately less convincing than the Christian redemption story. However, besides Fraser's critique, it is worth considering what other final metamorphosis Nietzsche might have invoked; what other incarnation captures as well the capacity to bring something new to the world than the child? Christianity does not have the monopoly on the redemptive image of the child, and it is not obvious that the third metamorphosis is a piece of "theological sampling" when you consider that the child is not just the image, but is the agent par excellence of new beginnings. And it is mothers, of course, who give those new beginnings in the originary morphosis of the human foetus and child. So in fact, we might contend that Christian eschatology and the theology of rebirth actually amounts to a complex and sophisticated process of *maternal* sampling. But it seems that to be born once, for philosophers, is never enough.

Like Plato and so many others before him, Nietzsche appropriates the acts and demands of maternity, the sacred Yes and first motion of life, for his own philosophy. While he claims to invert Platonism, he rehearses the notion of spiritual pregnancy and "male mothers" (Nietzsche and Williams 2001, 75) that only seems to reproduce the same conceit of a male facility for giving birth that we see in Plato. Irigaray writes quite explicitly of "the desire Nietzsche had to be a mother" (Froese 2001, 167), though in rejecting Platonism and Christianity the object of his creative labour could only be the self in its embodied and contingent finitude. If we are to be born only once, we can only affirm and possess our nativity such as it was in this world, together with the life it delivered us. As Frances Oppel explains:

> If 'escape is impossible' – and here 'escape' also means 'solution', or 'meaning' guaranteed by a comforting faith in a metaphysical 'real world' that 'justifies' human endeavor – then the only empowering move

possible, according to Nietzsche, is to affirm the negative, to trans-value its value, and to rejoice in the circumstances of one's fate as necessitating creativity. (1993, 94)

This move is to will the eternal return of the same, to be strong—or superhuman—enough to affirm the recurrence of one's fate, the self-same and self-created life, times without number. Nietzsche radicalises the Christian notion of eternity so that, in the absence of other worlds, we must be able to affirm our lives in this world over and over without end.

Empowered by Nietzsche's affirmation of eternal recurrence, Krzysztof Michalski writes that "In every moment my entire life is placed under a question mark: I am born anew, I become a *child*, with no past, no worries about the future, no wish to hold on to anything that passes" (2012, 200–201). Here, then, is the final metamorphosis that Nietzsche commends to us: forgetting our past and consciously propelling ourselves in the new beginning of every moment that will return to itself over again; empowered *in this world* by delivering and creating ourselves joyfully against the ostensibly negative thought that "the eternal hour-glass of existence will be turned again and again—and you with it, speck of dust!" (2001, 194). Without escape to a 'real world' or the meaning in linear time assured by a Creator, we are exposed to the existential questions of how to live now that "the horizon seems clear again" (199). Nietzsche exhorts an endless coming to life at each moment, a love for eternal return in circular time that frees up a disregard for the past and the gift of life that comes from a mother. As Kelly Oliver puts it, "The eternal return makes it possible to give one's life as a gift to oneself" (1995, 112). While Christ would have us born again but once into another world or realm, Nietzsche has it that we are born autochthonously again and again for eternity in this world. Fraser may have been wrong to claim for Christianity alone the redemptive image of the child, but he is right to challenge the forgetfulness required by Nietzsche's final metamorphosis that we should be born anew in each moment. He fails to elaborate that the object of that forgetting, as ever, is the mother.

Luce Irigaray reads Nietzsche's eternal recurrence as the outworking of his resentment of being born of woman. She addresses him directly in her book *Marine Lover of Friedrich Nietzsche*, interrogating his resistance to acknowledging and remembering the maternal gift of life and the marine element which, she explains in a 1981 interview *Le corps-à-corps avec la mère*, "evokes the amniotic fluids which thwart the eternal return" (quoted in Oppel 1993). Zarathustra entreats his followers to remain true to the earth, which he seems to identify as the mother to which we may return: "I want to become earth again, that I may have peace in her who bore me" (1961, 99). Irigaray challenges this easy peace and the erasure or forgetting of the sea, the ebb and flow of woman's time and women's bodies which give life: "You sought to become a child again, to climb ashore and drag your man's body once more. Why leave the sea? To carry a gift—of life. But it is to the earth that you preach fidelity. And forgetfulness of your birth" (1991, 12). It is only by forgetting the sea, the womb and fluid of his birth from a woman, that Nietzsche can carry the gift of life for himself, to credit himself with the labour of delivering his own life into each moment. Irigaray leaves us in no doubt as she addresses Nietzsche directly: "your whole will, your eternal recurrence, are these anything more than the dream of one who neither wants to have been born, nor to continue being born, at every instant, of a female other?" (1991, 26). Nietzsche wants to give his life to himself, to make himself a gift to himself, good enough to pass a test of circular time true to the earth; to claim himself as his own creation. But this can be achieved only by disowning his creation by another, by forgetting—whether wilfully or unconsciously—his generation and birth from the marine elements of the maternal body; by usurping the powers of procreation, pregnancy and birth.

Nietzsche foreshadows the end of a tradition of Western thought that has sought to identify meaning or salvation in a world beyond or more true than the one we inhabit, a world that has required a rebirth of human beings by labour of the mind or spirit. Yet, as Oliver explains, Nietzsche's attempt to redeem life after the death of God through an ethic of self-creation still appropriates the demand of rebirth common to that tradition: "in Nietzsche's texts, this birth is an auto-birth as self-overcoming. It is the birth of the self for the sake of the self; it is

the ultimate gift to oneself" (1995, 147). In making this gift to oneself, the gift of one's life given by the mother is devalued and eclipsed in
much the same way as in Platonic and Christian patterns of thought.[8]
Nietzsche's endeavor to revaluate all the values he inherited by way of
Platonism and Christianity could not finally extend to acknowledging
the neglected debt or gratitude to the maternal body. "The Übermensch
has no need for a mother; he gives birth to himself" (Oliver 1995, 146).

Conclusion: Goddesses and Monsters

We began the chapter by looking at two prominent bioethicists'
paper on the creation lottery and 'Frankenstein' reproductive technologies, and though it seems unlikely that we should have found our
way to Nietzsche, Harris and Savulescu's work does contain echoes of
Nietzsche's thought and language. For one thing, their flirtation with
the truth as goddess is prefigured by Nietzsche's supposition that truth
is a woman: "Aren't there reasons for suspecting that all philosophers, to
the extent that they have been dogmatists, have not really understood
women?" (1973, 31). Nietzsche figures his predecessors (and ultimately
perhaps himself) as failed suitors, spurned by the mysterious and elusive
womanly truth; Harris and Savulescu consider themselves less modestly
as having grasped the truth as woman, the profligate goddess creation
lottery. To be sure, Nietzsche is in agreement with them in their critique
of those who take (or worship) nature as a normative marker of proper
conduct:

> So you want to live "according to nature?" Oh, you noble Stoics, what a
> fraud is in this phrase! Imagine something like nature, profligate without
> measure, indifferent without measure, without purpose and regard, with-
> out mercy and justice, fertile and barren and uncertain at the same time,
> think of indifference itself as power – how could you live according to
> this indifference? (1973, 39)

Harris and Savulescu show that a moral commitment to this indifference in the lottery of natural reproduction implies approval of many

other interventions in the process of creating human beings. And further, the moral superiority of some reproductive technologies over the indifference of nature, to be colonised by the more just discernments of human judgment.[9] They are all uninterested in exploring a moral commitment to the sexed and oblational phenomenology of carrying and giving birth to others.

In spite of their reference to a goddess, Harris and Savulescu do not subscribe to the existence of other worlds or second births; rather, it is for the sake of this world and the welfare of those who enter it by birth that they would appropriate it by choosing children. To some extent, Nietzsche's philosophy of self-creation still provides the values that guide those selections, but rather than play God gifting life to oneself, selective reproduction allows the purposive and deliberate gift of life to others. The imperative is to bequeath a level of capability and autonomy to allow the created individuals the best chance of creating the lives they want and satisfying their preferences. To screen out, so far as practicable, any genetically inherited physical or behavioral obstacles to the free authorship of a given life, with the facility to choose from among as many choices as possible. However, in rendering natural reproduction a lottery, and thereby the maternal body at once a destructive and inadvertent maker of life, Harris and Savulescu's courtship of the human condition inherits a both Platonic and nihilistic disowning of any debt to the mother. Each of us exist only by chance, or may just as easily not have existed in the creation lottery of natural reproduction; for Harris and Savulescu, there is no significance in the manner in which we are borne into the world once the genetic roulette wheel has stopped spinning.

We have seen then, that in the practice of contemporary reproductive technologies the maternal body still signifies mortality and inferiority, only now it is for the prospect of its threat to and failure to deliver rather than through any deprecation of the birth of the biological self. The mother is still the vessel for the gestation of the child, but is now the object of suspicion not because she may give birth but because she may not—or not so well as she might or perhaps should. In a modern scientific society that has largely taken leave of spiritual worlds into which a second birth might deliver us, the onus is upon delivering the biological life of the other into this world as best we can, since that life is all that there is. As Godbout has written, "The child is the only transcendence we have left" (1998, 41). We are repairing to the existential priority of the first and only birth into this biological life, just as agents other than the mother can appropriate power over not just the birthing body but what it produces. In this way, male actors are able to participate in the remaining feat of transcendence; the act and relation of giving birth still eludes, but the selection or design of a given child comes within the gift of paternity, and we might consider that paternal gaze to extend to male clinicians, doctors, lawyers and indeed the state in negotiating reproductive decisions. Men can begin to determine the content of what is given, if not participate in the maternal context of the gift itself. Thus the content—who/what the child is (capable of)—becomes all-important; the context of maternal giving is scrubbed out as the mother is either merely the delivery system for the chosen child, or else an anarchic roulette in a reproductive lottery.

The enormity of the change that these increasingly selective forms of assisted reproduction portend, under the liberal eugenic endorsement of orthodox bioethics, starts to come into focus with the language of giftedness. For Harris and Savulescu, the maternal context of childbirth and child-rearing has never been sufficient to recognise giftedness in the life given against the brute chance of a child's existence.[10] In this, as we have seen, they follow a history in Western culture of denigrating or ignoring any significance in the fact that human beings are all given birth/life by women. Now that prospective parents or others are able to select in existential favour of particular children, their obligations to procreative beneficence are framed in terms of gift-giving virtues; once again, as Singer asks, "wouldn't we all prefer parents who try to make the gift as good as possible, rather than leaving everything to chance?" (2009, 279). Set against a history of renouncing the gift of birth/life from the maternal body, and subjecting that body to the sovereign gaze (and obstetrics) of a male order, this becomes the inevitable question with the suggested virtues that follow. Yet we have begun to see that this history may have been otherwise, and that some of the forces that made this history share a particularly masculine drive to deny or overcome the condition of being given birth by women.

Western patriarchal culture—and philosophy—has sought to decry any debt to the maternal body, through disparaging and appropriating its life-giving agency and seeking power over reproduction, lineage and the bodies of women. The characterisation of natural reproduction as little more than a creation lottery is a contemporary site of resistance to the acknowledgement of our maternal origins. This is not to say that the process of embryo creation in natural reproduction is *not* subject to chance, to the absolute and untold vagaries of conception, but to suggest that moral significance does not begin and end at that lottery. In the following chapter I will explore and develop the notion that life can rightly be considered a gift, given by mothers—a notion that our culture has taken such pains to resist. The idea that reproductive technologies allow prospective parents to switch simply from receiving the lives of unbidden children virtuously to giving bidden children well is complicit with the patriarchal denial that life is ever gifted by the mother. Rather, these technologies, re-evaluated in the acknowledgement of maternal giftedness, are allowing us to shift from one kind of parental giving to another, and it is in this light that a different story about them can be told, and the beneficence of selective reproduction might be more properly assessed.

Notes

1. Though note the zealotry of their own pretension to provide 'final' lessons from natural reproduction, which tends to suggest an unwillingness to accept further debate.
2. Having occasion to ask John Harris about this footnote at a conference, he recalled that it was Julian Savulescu's contribution. I have not had the opportunity to ask the same question of Mr Savulescu.
3. For further critique of Harris's blindness to sexual difference and the phenomenology of reproductive choice for women, see Kelly Oliver (2010) "Enhancing Evolution: Whose Body? Whose Choice?" *Southern Journal of Philosophy* 48: 74–96 (Oliver 2010).
4. See also Emily Martin's (1987) feminist analysis of the language used to describe menstruation as failed pregnancy and the gendered construction of sex cells in the process of reproduction in *The woman in the body: a cultural analysis of reproduction*, Boston: Beacon Press (Martin's 1987).

5. See also, for a thoroughgoing account of male pregnancy in Greek literature, David Leitao (2012) *The pregnant male as myth and metaphor in classical Greek literature*. New York: Cambridge University Press (Leitao 2012).
6. See also "On Truth and Lie in an Extra-Moral Sense" in Nietzsche, F. W. and W. Kaufmann (1971) *The portable Nietzsche*. London: Chatto and Windus (Nietzsche 1971).
7. More recent feminist theology has sought to revise and think beyond traditional patriarchal models of Christian thought, see for example: Grace Jantzen (1998) *Becoming Divine: Towards a Feminist Philosophy of Religion*, Manchester: Manchester University Press. Also, Rachel Muers (2007) "Feminist Theology as Practice of the Future." *Feminist Theology* 16(1): 110–127 (Jantzen 1998; Muers 2007).
8. There are notable examples of twentieth century thinkers appropriating the maternal metaphor for forms of self-genesis, Marx being a prominent example. For an account of Marx's language of birth, see Walker, M. B. (1998) *Philosophy and the Maternal Body: Reading Silence*. London; New York: Routledge. See also Parker, A. (2012) *The Theorist's Mother*. Durham: Duke University Press (Walker 1998; Parker 2012).
9. We might remark that this language of colonisation, introduced by Buchanan et al., also has rather unfortunate undertones and resonances with violent imperialism, figuring nature as an uncivilised land of opportunity.
10. Also see Kelly Oliver's deconstruction of Harris's chance vs. control, propagating "the fantasy of man's triumph over the forces of Mother Nature through which he gives birth to himself without the body of a woman/mother" in Oliver, K. (2013) *Technologies of life and death— From cloning to capital punishment*. New York: Fordham University Press, 78 (Oliver 2013).

References

Bachofen, Johann Jakob. *Myth, Religion, and Mother Right: Selected Writings of J.J. Bachofen*. Routledge & K. Paul, 1967.
Birke, Lynda I. A., Gail Vines, and Susan Himmelweit. *Tomorrow's Child: Reproductive Technologies in the 90s*. London: Virago, 1990.
Cavarero, Adriana. *In Spite of Plato: A Feminist Rewriting of Ancient Philosophy* [in Translation of: Nonostante Platone: figure femminili nella filosofia antica.]. Cambridge: Polity, 1995.

Corea, Gena. *The Mother Machine: Reproductive Technologies from Artificial Insemination to Artificical Wombs*. London: Women's Press, 1985.

De Graaf, Mia. "Babies Made without Mothers 'Will Come Sooner Than We Think', Leading Scientists Warm after Study Discovered How to Make Embryos from Skin Cells." *Daily Mail*, 2017.

Diprose, Rosalyn. *Corporeal Generosity: On Giving with Nietzsche, Merleau-Ponty, and Levinas*. Albany; [Great Britain]: State University of New York Press, 2002.

Fraser, Giles. *Redeeming Nietzsche: On the Piety of Unbelief*. London: Routledge, 2002.

Froese, Katrin. *Rousseau and Nietzsche: Toward an Aesthetic Morality*. Lanham, Md.: Lexington Books, 2001.

Godbout, Jacques T., and Alain Caille. *The World of the Gift* [in Translation of: L'esprit du don.]. Montreal; London: McGill-Queen's University Press, 1998.

Greer, Germaine. *The Whole Woman*. London: Doubleday, 1999.

Harris, John. *Enhancing Evolution: The Ethical Case for Making Better People*. Princeton, N.J.; Woodstock: Princeton University Press, 2007.

Heidegger, Martin, and David Farrell Krell. *Nietzsche* [in Translation of: Nietzsche.]. London: Routledge & Kegan Paul, 1979.

Heidegger, Martin, Gregory Fried, and Richard F. H. Polt. *Introduction to Metaphysics* [in Translated from the German.]. New Haven, Conn.; London: Yale University Press, 2000.

Holm, S. "The Creation Lottery and Method in Bioethics: A Comment on Savulescu and Harris." *Cambridge Quarterly of Healthcare Ethics* 13, no. 3 (2004): 283–287.

Irigaray, Luce. *Speculum of the Other Woman*. Ithaca, N.Y.: Cornell University Press, 1985.

Irigaray, Luce, and Friedrich Wilhelm Nietzsche. *Marine Lover of Friedrich Nietzsche; Translated by Gillian C. Gill*. New York: Columbia U.P., 1991.

Jantzen, Grace. *Becoming Divine: Towards a Feminist Philosophy of Religion*. Manchester Studies in Religion, Culture and Gender. Manchester U.P., 1998.

Kahn, Robbie Pfeufer. *Bearing Meaning: The Language of Birth*. Urbana, Ill.: University of Illinois Press, 1995.

Leitao, David D. *The Pregnant Male as Myth and Metaphor in Classical Greek Literature*. New York: Cambridge University Press, 2012.

Martin, Emily. *The Woman in the Body: A Cultural Analysis of Reproduction*. Boston: Beacon Press, 1987.

Michalski, Krzysztof. *The Flame of Eternity: An Interpretation of Nietzsche's Thought*. Princeton, N.J.: Princeton University Press, 2012.

Morgan, Lewis Henry. *Ancient Society*. Chicago: Kerr, 1877.

Muers, Rachel. "Feminist Theology as Practice of the Future." *Feminist Theology* 16, no. 1 (2007): 110–127.

Nietzsche, Friedrich Wilhelm, and Bernard Williams. *The Gay Science: With a Prelude in German Rhymes and an Appendix of Songs* [in Translated from the German.]. Cambridge: Cambridge University Press, 2001.

Nietzsche, Friedrich Wilhelm, and R. J. Hollingdale. *Beyond Good and Evil: Prelude to a Philosophy of the Future*. Penguin Classics. Harmondsworth: Penguin, 1973.

Nietzsche, Friedrich Wilhelm, and Walter Kaufmann. *The Portable Nietzsche*. Revised ed. London: Chatto and Windus, 1971.

Nietzsche, Friedrich Wilhelm, Reginald John Hollingdale, and Friedrich Wilhelm Single Works Nietzsche. *Thus Spoke Zarathustra ... Translated with an Introduction by R. J. Hollingdale*. Harmondsworth: Penguin Books, 1961.

Oliver, Kelly. "Enhancing Evolution: Whose Body? Whose Choice?" *Southern Journal of Philosophy* 48 (2010): 74–96.

———. *Technologies of Life and Death—from Cloning to Capital Punishment*. New York: Fordham University Press, 2013.

———. *Womanizing Nietzsche: Philosophy's Relation to the "Feminine"*. New York, London: Routledge, 1995.

Oppel, Frances. "'Speaking of Immemorial Waters' Irigaray with Nietzsche." In *Nietzsche, Feminism and Political Theory*, edited by Paul Patton. London: Routledge, 1993.

Parker, Andrew. *The Theorist's Mother*. Durham: Duke University Press, 2012.

Plato, Floyer Sydenham, Thomas Taylor, Plotinus, Guy Wyndham-Jones, and Tim Addey. *The Symposium*. 2nd ed. Westbury: Prometheus Trust, 2011.

Rich, Adrienne Cecile. *Of Woman Born: Motherhood as Experience and Institution*. Virago Press, 1977.

Savulescu, J., and J. Harris. "The Creation Lottery: Final Lessons from Natural Reproduction: Why Those Who Accept Natural Reproduction Should Accept Cloning and Other Frankenstein Reproductive Technologies." *Cambridge Quarterly of Healthcare Ethics* 13, no. 1 (2004): 90–95.

Singer, Peter. "Parental Choice and Human Improvement." Chap. 12 In *Human Enhancement*, edited by Julian Savulescu and Nick Bostrom, 277–291. Oxford: Oxford University Press, 2009.

Songe-Møller, Vigdis. *Philosophy without Women: The Birth of Sexism in Western Thought*. Athlone Contemporary European Thinkers. London; New York: Continuum, 2002.

Stonehouse, Julia. *Idols to Incubators: Reproduction Theory through the Ages*. London: Scarlet, 1994.

Throsby, Karen. *When IVF Fails: Feminism, Infertility and the Negotiation of Normality*. Basingstoke: Palgrave Macmillan, 2004.

Walker, Michelle Boulous. *Philosophy and the Maternal Body: Reading Silence*. London; New York: Routledge, 1998.

Whitehead, Alfred North. *Process and Reality. An Essay in Cosmology … Gifford Lectures … 1927–1928*. Cambridge: University Press, 1929.

4

The Maternal Gift of Life

Abstract This chapter seeks to recover a thought of maternal giving, beginning with Jacques Derrida's deconstruction of the gift and the idea of its distinctive realisation in self-sacrificial death as eluding the possibility of reciprocation. A similarly radical structure of giving occurs in birth, where the child is fundamentally unable to return the gifts of air, corporeality and time to the mother who gives. This points to plausible accounts both of the patriarchal fixation upon death, and of recent feminist and maternal thinking about ethics that identifies in our giftedness at birth a source of our ethical calling to responsibility for others.

Keywords Gift · Reciprocity · Maternity · Birth · Gratitude · Death

Some years ago, my grandmother bought me a set of kitchen knives as a Christmas gift. Upon opening the present, I was a little confused to find a penny attached to the gift she had given, which she implored me to return to her. She patiently explained that there's a superstition around giving gifts of knives, that they may, in time, sever the relationship between giver and recipient. To get around this there is a tradition, apparently widespread, of attaching a penny for the recipient to return

© The Author(s) 2017   **81**
S. Reader, *The Ethics of Choosing Children*, Palgrave Studies in Ethics
and Public Policy, DOI 10.1007/978-3-319-59864-2_4

to the giver, ostensibly turning the gift into a transaction and thereby avoiding the bad omen; if money immediately changes hands for the knife, it cannot be considered a gift, and that way the danger to the relationship is averted. In my family at least, it has been women taking care of gift-giving all along.

Human beings are complicated creatures: the gifts we give can be ill-conceived, inappropriate, or even nefarious; they can make and break relationships. For this reason alone, many would avoid the language of gift altogether, and resist drawing any conclusions from testing an act morally against its definition. However, we do offer up objects, acts and services to one another out of myriad desires and motivations, and we do recognise different moral qualities in the distinctions between these purposes. The language of gifts thus remains of interest insofar as it connotes a more virtuous mode of giving than others, and its deconstruction in recent continental thought opens up a plausible explanation for the patriarchal derogation or appropriation of maternal giving that I have suggested is perpetuated by the narrative of Procreative Beneficence.

There is vast and diverse treatment of the gift, spanning decades and disciplines, and therefore I can't adequately account for all of it within this short book. Much of this literature originates in the field of anthropology, and particularly the function and facility of gifts in ancient or primitive societies before modern ideas of capital and commodity transactions. Most literature on the gift takes inspiration from Mauss's seminal 1925 study *The Gift*, which was an ethnological, historical and sociological enquiry into the workings of the gift in pre-market systems of exchange. Mauss discerned in the practices of Pacific islanders and American natives an elaborate circulation of gifts that engaged the honour of both giver and recipient, so that although the gift might appear free and unqualified it is in fact tacitly constrained. The theory goes that a certain economy inheres in the circulation of objects as gifts prior to the advent of market economies: there is an obligation to give in the first instance to establish or preserve authority and demonstrate honour; an obligation to receive a gift well, showing that one is grateful and unafraid of reciprocating; and an obligation to reciprocate since one is made inferior by the acceptance of a gift without the intention or capacity to return the gesture. Thus gifts seem to function under the ritual pretence of unilateral generosity whereas in fact they are exchanges according to well-established social rules and conventions. There is much that can be said of Mauss's analysis, particularly by way of postcolonial and feminist critiques of a tendency to overlay or corroborate certain of our own presuppositions and biases in other historic cultures.[1] For our purposes, it is the connection between morality and the gift that is of interest, insofar as we tend to discern greater good or virtue in the givers who, like my grandmother, honour an idea of giving that is unconcerned by reciprocation. It is this discernment, pressed and radicalised perhaps most famously by Jacques Derrida's reading of Mauss, that can tell us more about the beneficence bound up in giving birth.

Reciprocity and the Problem of Recognition

Derrida's reading of Mauss takes up the dynamic of gift-giving as indicative of the problem of ethics itself. He reads Mauss's analysis of the obligations and returns incumbent upon the gift as fatal to the very idea of being ethical at all, stretching the criteria of what reciprocation consists in to consider the reimbursement effected by any compensation for the act of giving. For there to be a gift, Derrida argues, "there must be no reciprocity, return, exchange, countergift, or debt. If the other gives me back or owes me or has to give me back what I give him or her, there will not have been a gift, whether this restitution is immediate or whether it is programmed by a complex circulation of a long-term deferral or difference" (1992). By these exacting standards, we interrogate the giftedness of even birthday or Christmas presents and find that a basic economy of exchange inheres in this tradition; we do not demand money for our gifts, but the expectation is that a gift will be received in return or at some later juncture. Without reciprocation, gift relations do not last very long. We might consider that gifts or donations given anonymously to strangers are more properly examples of genuine gifts that defy reciprocation. Yet Derrida suggests that a kind of return occurs in the very cognisance of the gesture. When one gives, one receives the satisfaction of the good that one has gifted, in the knowledge that one's gift has been received, registered and may be remembered. Derrida claims radically that recognition itself acts as a symbolic equivalent of a return, a form of reciprocation even when a return gift is not forthcoming.

In his reading of Mauss, then, the insight that gifts function according to a principle of reciprocity demonstrates for Derrida that "every gift is caught in the round or the contract of usury ... there is no gift without bond, without bind, without obligation or ligature" (26–27). Based on an understanding of the gift as that which is irreducible to economic calculation, Derrida teases out the suspicion of the impossibility of a consistent discourse on the gift, the impossibility of the gift itself. "One could go so far as to say that a work as monumental as Marcel Mauss's *The Gift* speaks of everything but the gift: It deals with economy, exchange, contract (*do ut des*), it speaks of raising the stakes, sacrifice, gift and counter-gift—in short, everything that in the thing itself impels the gift and the annulment of the gift" (24). For Derrida, the gift is a paradox in the sense that the recognition of a gift *qua* gift is a necessary and sufficient condition of its annulment. As soon as an object or gesture appears to be a gift it is absorbed into the closeted economy that Mauss had brought to light. Most obviously of course, the recipient or beneficiary should be oblivious to the gift since, as we have seen, it invites reciprocation, even if by way of mere gratitude. Yet for Derrida the apprehension of the gift by the giver alone acknowledges itself, registers its gratuity and largesse, and by virtue of this offers itself a return of sorts.

The problem that this creates for ethics is that the goodness of a gesture is thought to be compromised by self-interest, and "the simple consciousness of the gift right away sends itself back the gratifying image of goodness or generosity, of the giving being who, knowing itself to be such, recognises itself in a circular, specular fashion" (23). Gift exchange, as Mauss sought to explain, reduces to a kind of economy where, in fact, we are always conscious of the obligations our giving imposes upon others and the honour it imposes upon ourselves. So, then, if we are conscious of the goodness of a gesture, which we must be for us to consider it moral, then that consciousness itself impresses upon us the gratifying knowledge that serves as a kind of return and compromises the purity of the gesture. Insofar as we realise our own beneficence in the gift, there is self-interest in the giving of it which precludes the possibility of it being truly ethical.

This requirement that the ethical gesture be disinterested, that the moral content of an act is contaminated by the thought of a return, is often regarded as an undeniably and historically Christian concept (Hyman 2004, 35). We should act out the good for its own sake, and the mode of ascetic or even violent self-sacrifice exemplified in Christ and the requirements of Christian life tell of a commitment to the good made in spite of fear and regardless of favour. Philosophically, the ethical ideal of disinterestedness follows from a Kantian epistemology which demands that the moral law must be observed for its own sake and not be prompted by partiality to favourable consequences or anticipation of a reward. Everything must be "disinterested and grounded only on duty, without being based on fear or hope as incentives, which, if they become principles, would destroy the entire moral worth of the actions" (Kant and Gregor 1997, 108). If our acts so much as entertain some personal profit or diversion of harm then they cannot be reckoned to be moral; the interests of the subject compromise the morality of their acts, so the subject must be denied or negated to achieve true morality. For Derrida, we have seen, it is ultimately not possible to negate the self so entirely that no return is conceived for the ethical act, or gift, if it is realised as such.

This leads to a school of thought that makes our mortality the defining predicament for the human subject, and the condition of being vulnerable to death the very circumstance that makes it possible to be ethical at all.[2] The closest we can get to the good is in literally sacrificing the self for others since a sacrificial self-offering can expect no personal benefit in return. As John Milbank (1999) explains, "At the limit, the ethical agent might die for the vulnerable other person. This readiness to die alone guarantees the ultimate disinterest of his ethical gesture, since it would seem that a good one is prepared to die for cannot be the secret vehicle of one's own power or (presently enjoyed) glory" (33). There is, within the structure of the event of self-sacrifice, an erasure of the subject who gives (their life) that absolutely eliminates the possibility that they will benefit from the act. Since upon the destruction of one's own life there can be no return, it is one's death which, in the cause of another, might be supposed to be the purest gift that one can give. In self-sacrifice the good exceeds any self-interest as the possibility of a return is extinguished along with the life given; the gesture itself removes from the equation the giver to whom the obligation or recognition is incurred. As the giver is removed from existence by their own gift, it is inconceivable—unless one supposes that there is an afterlife— that they can be rewarded or benefit from some manner of reciprocation. Only the erasure of the giver from the existential equation breaks the circle and guarantees that the gift is not compromised by the possibility of a return: "Only the dead person, on this account, only the subject who has passed beyond subjectivity, can be a true giver" (Milbank 2003, 156).

It is also death, so long the preoccupation of philosophers, which is therefore said to orient our subjectivity and sense of responsibility to others. It is a common enough platitude, of course, to assert that there is nothing in the world of which we can be certain but death and taxes, a phrase variously attributed to Daniel Defoe, Benjamin Franklin and Mark Twain. However, there is some philosophical work that follows from this idea which suggests we value the provision of that certainty in death as that which conditions the demand of the ethical. Of course death underpins our vulnerability as subjects ourselves, and the appeals we make to others, but as Milbank explains "only the capacity of the ethical subject to respond to the needy person, if necessary with his own death, guarantees his deed as truly ethical, as a truly disinterested gift" (2003, 154). The category of mortality is therefore valorised as the only condition about which we may be certain and the only condition about which the notion of giftedness can be countenanced in light of the strict disinterestedness required of the ethical gesture. Indeed, a consistent and thorough commitment to disinterestedness of course precludes belief in immortality; death remains the preoccupation of philosophy not because it releases the soul from the body or into a life eternal, but precisely because it does not, and by that token is supposed to be the singular calling to ethical responsibility.

It can be no coincidence that Western thought has been so obsessed by death and the idea that the giving of one's death represents the apotheosis of the gift.[3] Of course, to insist upon the necessity of dis-interestedness in ethics, makes impossible and suicidal demands upon the ethical subject. As Hyman (2004) explains, "in so far as one affirms (one's own) life, one sacrifices disinterestedness and in so far as one affirms disinterestedness, one sacrifices (one's own) life. Death, it seems, is a necessary precondition for a realisation of true disinterestedness" (44). That it is not attainable, except perhaps in death, does not mean that disinterestedness is not a genuine or laudable ideal any more than it means that gifts cannot therefore exist. Derrida maintained that while the gift cannot be known as such, it can yet be thought of; thinking through its impossibility opens up insights into the limits of knowledge and human economies of exchange.[4] I offer no judgement, for now, on the value of the imperative to disinterestedness, but only seek to estab-lish that we inherit an intellectual tradition that commends it to us, and what Milbank calls a "recent consensus" of ethical thought that iden-tifies and celebrates its precondition in the gift of self-sacrificial death (2003, 154–155). That is, there is in the dynamic of dying for another the unique erasure of the possibility of a return; self-sacrificial death is profoundly irreducible to contract—and therefore ethical—since it is a gift that takes the giver from life.

The Gift of Life

While being vulnerable to death secures the possibility of true disin-terestedness, being vulnerable to giving birth seems to compromise it. Yet I want to claim that this is precisely the reason that we should consider the dynamic of giving birth as being quite as morally signifi-cant as that of self-sacrifice, or the gift of death. The philosopher John Protevi (2001) makes a similar move, and, despite the grandiosity of the phrase, writes more boldly of the gift of life. While death is the appar-ent preoccupation of Derrida in his writing on the gift, Protevi points out the specificity of gifting life—of being gifted life—in relation to Derrida's analysis of giftedness. Lisa Guenther (2006) also makes a simi-lar case, and reads Levinas to help us consider the meaning of birth as the gift of the other. I will follow both writers in seeking to recover an understanding of birth or life as gift in order to consider the moral significance of the shift in the practice of giving life that has been occasioned by selective reproductive technologies.

We can make the gift of death because in extinguishing the self that gives, the possibility of recognition between giver and the other for whom death is given is ruled out. The gift itself erases the giver, and it is this existential erasure that has preoccupied the philosophical contemplation of finitude. Yet the event of birth as the root of our finitude displays a similarly arresting dynamic in the existential novelty of the newborn as receiver of life from another, creating the recipient in the act of the gift. The possibility of recognition between giver and the other to whom life is given in birth is actually created. While death removes for the giver the possibility of self-interest, birth creates the possibility of interest—or, to put it another way, the impossibility of disinterestedness. The recognition of life as gift "frees a thought of maternal production of life" (Protevi 2001, 82) that has hitherto been appropriated by monogenetic and Aristotelian notions of a dominant male principle working upon the passive material of the mother in reproduction, or else some theology of divine giftedness in the life of every child. Some appreciation of giftedness has always been present in procreation, but also a fierce resistance to acknowledging mothers or the maternal body as the source of the gift.

To be sure, while we should no longer countenance the Aristotelian notion that procreation is the outcome of "the spermatic motions of the father" (79), sexual difference has been the premise of natural reproduction to date, so some might prefer to recognise this as the source of the gift rather than the mother alone. That is, there is inscribed in a natural or given order a complementarity between the sexes in their biological difference that allows for the creation of new lives. Theologian John Milbank (2003), for instance, arrives at such a view following a reading of Irigaray, but there are dangers inherent in extrapolating from difference to complementarity, such as his offering the following observations: "men are more nomadic, direct, abstractive and forceful, women are more settled, subtle, particularising and beautiful" (207). Note that Milbank here contrasts his male characteristics with their opposites in his rendering of women, e.g. nomadic/settled; direct/subtle;

abstractive/particularising; except in the final trait where he describes men as forceful, and women beautiful! If men are indeed more forceful in some mode that women are not, Milbank cannot quite bring himself to express the complementary characteristic of women, which would presumably be something like passivity or docility. While this would seem to follow from his formulation, it would betray the more impolitic nature of his analysis; his solution, citing the greater beauty of women, is fantastically patronising. He ignores the possibility of women (or non-heterosexual men) as desiring subjects, and, like Harris and Savulescu, positions women in the admiring or desiring gaze of the more forceful man.

So there are real dangers in taking a view of complementarity that somehow follows from sexual difference, though the question of sexual difference—or the place of the father—as giving the gift of life remains.[5] I don't want to discount or ignore sexual difference in the present necessity of procreative sex, but suggest that there are good reasons why we might privilege the biological role of the mother in reproduction as doing the work of gifting new life—and reasons that go some way to explaining the effacement of the maternal body in philosophy. In thinking sexual difference a risk is that we don't think beyond sex itself; Irigaray reminds us in her reading of Heidegger that "long before the newborn takes a first breath of air on its own, the pregnant woman has already been breathing for the foetus through the mediation of a placenta that both connects them and keeps them distinct" (Guenther 2008, 100). While birth involves beginning to breathe on one's own the same air among others, the foetus is sustained by the breath and oxygen of its mother until it is born. Conception might be the result of sexual difference, but gestation relies absolutely upon the hospitality of the maternal body—which is shared or gifted by the mother alone. As Cimitile explains, in Irigaray's reading, "Heidegger covers over the material transcendental condition for Being—the first beginning, the gift of the mother" (2013, 121). This is not a gift we can remember, but Irigaray reads a pathological forgetfulness of this gift—a 'forgetting of air'—in Heidegger's consideration of human finitude and the tradition of thought he passes on (Irigaray 1999). Whatever else we think, say or become in life, there is a priority of birth since even if one orients one's being toward death it presupposes an initial gift of breath which comes from the mother.[6]

Here we get to the crux of the gift of birth or life—which gives air, just as it gives time for life to an other: the maternal gift cannot be captured within an economy of exchange as it is given in a profound asymmetry that resists reciprocation. As Irigaray writes: "She gives first. She gives the possibility of that beginning from which the whole of man will be constituted. This gift is received with no possibility of a return. He cannot pay her back in kind" (1999, 28). Just as the gift of death, in giving up one's own life, precludes the possibility of a return, the gift of life importantly resists reciprocation since one cannot return to the mother the unique and uniquely possessed life she already has. In this way, it shares a structure of non-return with the gift of death: no matter what manner of hospitality, gratitude, or gifts I might offer to my mother in the course of my life, I can never give life itself to my mother as she gave mine to me. I might be in a position to provide for her more life—by way of an organ donation, say—but this would be an extension of simply more of the same life she already possesses. Indeed then, as Protevi suggests, "the gift of life, precisely in this excess of gift over symbolic exchange, is, if not quite the paradoxically self-effacing 'gift itself', perhaps the closest to the gift of all our idioms. If there were to be a gift, it would be the gift of life" (78).

As we have seen, both death and birth have been supposed to be the gift par excellence due to an ontological impossibility of contract whereby death extinguishes the giver and birth animates the gift. As Guenther (2006) explains, the gift of birth is "unique in that it gives rise to its own recipient; the child who receives the gift of birth exists only thanks to this gift. In this sense, birth is not only given to me; it also gives me" (2). Where self-sacrifice is unique in being given from me and also giving up my existence, my birth is given to me and also gives me existence in an act or dynamic of what Protevi calls "irreducible excess" (76). Where other exchanges may be supposed reducible to speculative returns, gifts of life and death generate or annihilate the subjects giving or receiving gifts and therefore exceed the possibility of contract in radical, existentially-affective ways. Yet unlike the gift of death, life can only yet be given by a woman, and while death/breath might be given for more life for another, birth gives brand new life to somebody yet to breathe/live alone.

The uniqueness of this gift also explains the slippage that emerges between terms like birth, life, air and breath—the kind of slippage that is acknowledged in the writing of thinkers like Derrida and Irigaray, but that no doubt makes the concept intolerable for more analytic philosophers. To be sure, under other circumstances such terms denote very different things, but acknowledging the various elements that are given or created with a life simply involves working and writing with these entanglements. Nevertheless, it is important to register just what it is that is given in the gift of birth, just what the gift gives. As Irigaray has shown us, the mother gives air and breath to the natal in the womb; the act of conception gives potential form, sex and a genetic makeup that is nurtured and carried in and by the body of the pregnant woman. Of course a particular home, life, culture and history are also part of what is given to a child, and we will consider this further in the following chapter. Yet what is also given perhaps more fundamentally than anything else is time. "Quite literally, birth "gives" what no one—neither the mother nor the child—"has" in the sense of possessing something; it brings forth the time of the other. This gift of birth is not a simple transfer of existence from one subject to another; rather, it gives rise to the Other to whom it will have given birth" (Guenther 2006, 52). In this sense, we see again the way in which the gift of life exceeds the circularity of exchange whereby a new existent emerges in newly given time.

Although Derrida gives extensive treatment to the gift and time, he resists acknowledging the time that is gifted by mothers—or at least that this is philosophically or ethically significant. This seems like an aberration, for he asserts that "the difference between a gift and every other operation of pure and simple exchange is that the gift gives time. *There where there is gift, there is time*" (1992, 41). As Protevi explains, Derrida's refusal to explicitly render giftedness from the maternal body is a resistance to the temptation of thinking such a gift as an ordering principle or "transcendental unifying given" (83) for the human condition. Derrida backs away from all such traditions of identifying or valorising such givens, but having made the gift so central to his thought on time won't quite follow it back to the maternal given that always fundamentally gives time for the other. Although she is scarcely part of such a tradition, Derrida tellingly includes the mother in his reckoning: "This great transcendentalist tradition can inscribe the transcendental given in the present in general (the present appearing of that which appears in the light, or else created being, the originary given of a gift which comes down to and comes back to Nature, Being, God, the Father—or the Mother)" (1992, 53). Ultimately, Kelly Oliver argues, Derrida's blind spot concerning the maternal body arises because he associates it with that transcendentalist tradition of locating the source of our given-ness. Yet, as Oliver wonders, "doesn't the acknowledgement of the debt of life always bring with it the danger of making the mother into a god?" (1997, 67).

Not only does the mother give time to the other for its own life, but she also gives time of her own in pregnancy, birthing and (often) breastfeeding her child. These are also acts of giving from/of her body; what Myra Hird, following Rosalyn Diprose, has called the corporeal generosity of maternity (2007). Diprose's work draws attention to the fact that every life is dependent upon a range of material openings to other lives that might otherwise be inscribed as debts to other bodies. That is, our dependence on other bodies relies on a generosity that is the "nonvolitional, intercorporeal production of identity and difference that precedes and exceeds both contractual relations between individuals and the practices of self-transformation" (2002, 75). Hird takes the exchanges of physical matter that occur during maternity as exemplary of the "excess, unknowability and openness" of corporeal generosity, citing genetic, placental and blood gifting, along with breastfeeding as the material, maternal gifts that we are absolutely dependent upon for our lives. "This process (life) is the first gift and first debt any living organism encounters, and cannot be shoehorned to resemble a familiar closed economic system" (14). The uniqueness of this gifting, or what is given, situates our lives as received in incalculable excess from other bodies, and most conspicuously from the bodies of our mothers.

The fact that the inceptive gifts of air, time and corporeality can only be given by women, might explain their forgetting in a history of philosophy dominated by men, who have concentrated the contemplation of finitude upon the death that we can all possess as our own rather than the birth that is given to us as oblivious and unremembering subjects by women alone. As Seyla Benhabib (1992) succinctly puts it: "The denial of being born of woman frees the male ego from the most natural and basic bond of dependence" (156). Not only is this an infantile physical dependence at the beginning of life, but also a psychological consciousness of an elusive, contingent—but essential—origin in the gift of a mother. We have already seen the formative philosophical roots of this denial in the Platonic ideas discussed in the last chapter, but the 'mother trouble' of Derrida and other more recent thinkers is also well accounted for elsewhere (Parker 2012). Among notable variations on diagnosing such trouble with the maternal is Julia Kristeva's category of abjection, a term she coined to articulate psychic and cultural aversion to phenomena and experiences that threaten the subject. The abject describes the events and encounters that disrupt the integrity of the conscious ego, that throw into doubt the distinction between self and other, prompting disgust or violent repulsion of that which jeopardises it. Seminal to the phenomena of abjection, for Kristeva, is the experience of birth itself: "Abjection preserves what existed in the archaism of pre-objectal relationship, in the immemorial violence with which a body becomes separated from another body in order to be" (1982, 10). That other body is always a maternal body, giver of "a repulsive gift" (9) that opposes and challenges my detachment and autonomy, reminding us that we were first expelled from and subject to the body of another. Kristeva posits this maternal abjection as a necessary psychological mechanism to become a subject in one's own right, but it is an abjection that is all too easily folded into encultured hostility to the maternal body and, by extension, women themselves.[7]

The more psychoanalytic accounts of rejection—or abjection—of the maternal body that gives us life, are not inconsistent with our analysis identifying the patriarchal root of that disavowal in the arrears to a gift that cannot be returned in kind, a debt to the maternal body that cannot be repaid or recreated. The aversion is perhaps not so visceral or physical, but neither is it calculated exactly; rather, learning the practices of human exchange and reciprocity that we are accustomed to, it dawns on us that no return for the originary gift of our lives is possible—and for men, that the power to generate other bodies alike is beyond them. As Irigaray put it to Nietzsche: "And that the other has given you what escapes your creation is the source of your highest resentiment" (1991, 42). While women might go on to gift breath, time and new lives as mothers themselves, men learn that they might only give the last breath and time of their own lives so it is this, historically, to which they attach meaning and value—to life which it is in their gift to give away, rather than that which it cannot be in their gift to give. Giftedness—that is, our being-gifted and the exclusive gifts of maternity—therefore becomes one plausible explanation for the inherited suspicions of the maternal body, and for patriarchal denials of equality for women in intellectual and faith traditions. Rather than admit to a fundamental and unpayable ontological reliance upon the bodies of birthing women, philosophers have sought to intellectually write off the 'debt' along with any significance in being given.

Being-from-Others

The development of a philosophy of the maternal, or maternal thinking (Ruddick 1989), has grown up in response to the relatively recent feminist critiques of patriarchal Western thought, showing how the voices and experiences of women have been silenced. These critiques are now fairly well rehearsed—though yet to form part of the basic curriculum in most cases—and various attempts have been made to move the discussion on from unpicking the silences to voicing the forgotten concerns and perspectives of women who give birth.[8] It seems plausible to suggest that a cause of these countless denials has been the association of the feminine with the maternal, and the radical dismissals and appropriations of birth by male philosophers points to a profound aversion to the idea of a contingent entry to the world, *inter faeces et urinam* and—which is worse—from a woman. Michelle Boulous Walker (1998), for example, has provided a persuasive argument for the claim that Western philosophy has silenced women most effectively by their association with maternity. With maternity the root of this aversion, or silencing, slowly attention is being given over to recover a thought of just what kind of gift birth might be, and the significance it holds for the contemplation of finitude and morality.

One helpful, if unorthodox, way into the discussion is to consider the insights of the German sociologist Georg Simmel, whose essay *Faithfulness and Gratitude* (1950) helps us prise open a problematic in the Maussian analysis of gift exchange, which accounts for giving, receiving and reciprocation but not the lingering obligation that the original gift imposes upon the other. That is to say, that one gives before the other receives and reciprocates, and to the voluntariness of that first gift the receiver can never be equal: the receiver of the gift may choose to reciprocate but they are conscripted into the exchange by the giver in the first place. When that gift also conscripts the recipient into being itself, creating a unique and singular life that was not there before, the asymmetry is absolute. The initial obligation into being with the other was pressed upon them at the gift of the other. As Simmel observes, there is a deep incommensurability when "once we have received something good from another person, once he has preceded us with his action, we no longer can make up for it completely, no matter how much our own return gift or service may objectively or legally surpass his own. The reason is that this gift, because it was first, has a voluntary character which no return gift can have" (392). The initial gift is given with an unbidden volition that a return gift cannot possess because it is not occasioned by gratitude or an obligation prevailed by the receipt of a prior gift from the other. Not only is the breath and time of life not returnable in kind to the mother who gives, but nor can the indebtedness be lifted from the originary gift of breath that also gave me my existence.

This added incommensurability may explain not only the trouble for (male) philosophers in coming to terms with their maternal origins, but also opens up the possibility of an approach to the gift, and economy more generally, that does not seek to discern a subject's inferiority or superiority in relation to an exchange. The gift of life would seem to resist the possibility of reciprocation, leaving the recipient of such a gift—that is all of us—indebted fundamentally to the mothers who gave us. According to the anthropological and philosophical interrogations of exchange we have seen, that would appear to leave us all in a state of inferiority, for, as Mauss observed, "the unreciprocated gift still makes the person who has accepted it inferior" (2002, 83). It is according to this logic that acknowledgement of the maternal gift is so vexatious for male philosophers, for the asymmetry creates an insistent indebtedness that holds no prospect of redress. Here we see how religious notions of gift work so well, in holding us answerable to a Creator or saviour to whom we owe our lives in a gift we cannot hope to reciprocate, but for whom we can strive to live in gratitude and humility. But would the maternal gift of life work in the same way? Simmel hints at a departure from the more economistic evaluations of being in relation to incommensurable giving in affirming the persistence of a disposition of gratitude in that relation as essential to social cohesion. "One might say that here gratitude actually consists, not in the return of a gift, but in the consciousness that it cannot be returned, that there is something which places the receiver into a certain permanent position with respect to the giver, and makes him dimly envisage the inner infinity of a relation that can neither be exhausted nor realised by any finite return gift or other activity" (392). Simmel's rather poetic expression of this relation prefigures subsequent attempts to write the maternal gift, as inculcating the ethical, back into an account of moral responsibility.[9]

Simmel writes of gratitude as the "inner infinity" of human relations under the realisation that we relate not as autonomous self-possessing individuals but as subjects to unending debts to one another; he refers to gratitude as "the moral memory of mankind" (388). Some of this is echoed by the philosopher Emmanuel Levinas (1969), who used similarly religious language and made a similar suggestion in asserting the idea of ethical responsibility as preceding, or perhaps pre-empting, the intellectual will of the subject. Levinas challenged the long-founded philosophical assumption that experience can be subordinated to a totalising ontology so that moral responsibility can be reckoned to follow from some definitive knowledge or truth. He showed that the premise or pretence of supposing to comprehend others is always a reduction of their alterity, or infinity; a subjection of the phenomenological encounter with the living presence of another to ontological speculation. Against this, Levinas argued that there is a structure of responsibility edified in human relations that precedes these more calculated forms of relating, recalling an ethics as first philosophy, or as prior to philosophy. Like Simmel's moral memory of gratitude to others, Levinas posits a pre-autonomous calling to responsibility for the other drawn not from an intellectual estimation or respect for their personhood, or as the out-working of a moral law or disposition, but as the traumatic experience of obligation demanded by encounter with the other. As Diprose (2002) explains, Levinas's work "lends itself to a philosophy of the gift, insofar as he bases a sociality that does not absorb difference on giving to the other without expectation of return. Subjectivity, for Levinas, is the pas-sivity of exposure to another, a giving of oneself without choice, a move-ment toward another arising from a disturbance of the self provoked by the other's alterity" (13–14). That is, it places at the heart of responsibil-ity the absolute necessity of the gift, the response of the subject to the infinite and unmeetable demand of the other.

The ethical plays out for Levinas as a radical hospitality to the other that is prior to any calculation of reciprocity or consideration of how or whether they merit it; the ethical is being hospitable. It is a welcoming subjectivity that does not thematise the other, does not stall to divine or negotiate an economy of exchange or vouchsafe some return for its generosity; it is exposure to welcome the other without the fulfilment of even the smallest proviso. It is an exposure that gives me the capacity to give, offering "powers of welcome, of gift, of full hands, of hospital-ity" (1969, 51). The paradigm of this relation for Levinas, the model for ethical responsibility as at once host and hostage to another, is in the munificent welcome of the maternal body which provides the first encounter, which gives the first gift of life. In *Otherwise than Being* (1981), Levinas suggests that ethical responsibility obliges us to bear the other "like a maternal body" (67) even though they may be a stranger whom I have "neither conceived nor given birth" (91). To be sure, this reference to the maternal body has invited much criticism from feminist thinkers all too aware that such an invocation of maternity as extreme generosity simply romanticises the trope of motherhood and reinforces the inveterate self-sacrificial expectations imposed upon women over the centuries.[10] As Jantzen observes, to "women who are responsible for caring for others, whether children, the elderly, or even students, it is no news that intersubjectivity is not reciprocal" (1998, 244). However, for Lisa Guenther, Levinas's recognition of the unilateral generosity in maternity points to an ethics that acknowledges birth as a gift that establishes the subjectivity it welcomes to the world.

In her brilliant critical rethinking of Levinas, mindful of all the feminist qualifications that should be made in reading him, Guenther takes as a touchstone for her inquiry his phrase "like a maternal body," emphasising the word "like" to open up a gap between maternity as a biological fact and as an ethical response:

> If to be born is to be given the imperative to bear even the stranger whom I "have neither conceived nor given birth to" (Num. 11:12), as Levinas suggests, then both men and women are commended to become like a maternal body for the Other, whether or not they give birth in a biological sense. I interpret this ethical imperative as a command even for the virile, autonomous self to be feminised and maternalised in his encounter with the Other. (2006, 7)

That is, being *like* a maternal body, whether one physically is or not, is to envisage and aspire to the same unconditional welcoming that the maternal body is subject to in bringing forth life. Picking up the pieces from Levinas's ethics and critical feminist readings of philosophy, Guenther moves the thinking on from considering mothering as gifting, to thinking embodied being as gifted in such a way as to expose us all to the ethical demand of being (gifted) from others and the responsibilities it creates, speaking of "our reception of existence as a gift that can never be reclaimed as a possession or choice but that precisely as such demands a response and perhaps even responsibility" (2008, 113). The gift is only one we can reproduce, not just in giving new life as parents ourselves but in all the acts of generosity we perform for others who are equally unchosen and gifted like us.

This call to be feminised echoes Hélène Cixous's identification of libidinal economies, which she appoints as masculine and feminine, although not essential to either sex. For Cixous (1986), the drive to secure or discern a return on one's investments and realise reciprocity is compelled by the "typically masculine ... fear of expropriation, of separation, of losing the attribute" (80). She wants to draw a distinction between libidinal economies that give and respond differently to gifts. A masculine or closed economy recoups its losses in a circle of debt and repayment, and deciphers giving as always a kind of indebting. "For the moment you receive something you are effectively 'open' to the other, and if you are a man you have only one wish, and that is hastily to return the gift, to break the circuit of an exchange that could have no end ... to be nobody's child, to owe no one a thing" (48). It is escape or avoidance of this openness to the (m)other, this vulnerability, that has driven masculine exchange practices, and indeed the matricides enacted in the various second births of philosophy, religion and finally the autonomous subject of liberal individualism, who "maintains the fiction of self-possession by imagining his birth as an autochthonous miracle, in which he springs from the earth fully grown and ready to take on the world" (Guenther 2006, 10). Cixous wants to escape this fiction, and the mercantile calculations of closed economies, to open up a thought of ethics as feminine exchange practices that can live with the gift.

What Cixous wants to characterise as feminine economy is differentiated by more open-ended expenditure, by truly generous giving that is not bent upon recuperation or reciprocity, and where "all the difference lies in the why and how of the gift, in the values that the gesture of giving affirms" (1986, 87). What is given is not measured, enumerated, remembered as a function of deferred exchange, but rather "without calculating, without hesitating, but believing, taking everything as far as it goes, giving everything, renouncing all security—spending without a return" (74). While this difference between masculine and feminine economy is not essential, such differences are enacted and inscribed in material bodies, and the maternal body is exemplary of a feminine economy in giving air, corporeality and time to another without the possibility of recuperation or reciprocity; it is the ground of another possibility of conceiving the human exchange of goods and acts as gifts rather than credits and debits upon the self-possession of discrete individuals.[11] The realisation of giftedness, of being subject to and in giftedness, frees us to act outside of that logic in a maternalised or feminised economy of encounter with the other. As Lisa Guenther suggests, "giving thanks for my birth does not amount to the repayment of a debt; rather it asks me to extend generosity to a stranger, to embody the gestures of feminine welcome for an Other who "deserves" my hospitality just as little as I have deserved the gift of birth" (2006, 56). What seems to be an essential premise of this gratitude that implores us to emulate a maternal welcome is that the gift of birth is undeserved, that the demand from the other is entertained beyond or prior to the contemplation of desert, just as—or because—we are all in receipt of a unique, unwarranted and untradeable gift in life.

The speculations on a feminist philosophy of the maternal briefly reviewed in this chapter begin to reconnect birth as gift with wider questions of ethical responsibility itself, including our collective responsibility to future generations. Rachel Muers (2008) has contributed importantly to this question, expanding the scope of some of the maternal thinking we have considered to the lives of communities and societies, to propose a notion of the social maternal. This idea convincingly draws every member of a society, mother or not, into a mode of relating to the future that recognises the fluidity and interdependence of self and other, past and future. Even if we are not mothers ourselves, the recognition of our emergence from and with maternal bodies bids us mother the future—socially, politically—*like* a maternal body. Importantly, this emphasises that "a crucial aspect of taking responsibility for future generations is forming the context in which their life takes shape" (135). That is, we all take responsibility not just for the lives and bodies created or selected to go well in the world, but also for the environment we create for those lives to go well in. Mothering the future, we take responsibility in giving first to those who will be born into and receive the time and environment of their lives from us; giving that future like a maternal body, then, perhaps means giving hospitality like a maternal body—open to the disruption of our own subjectivity and identity by the unknown, and letting go of the anticipation of reciprocity. In the final chapter, I will go on to consider what this might mean for a politics of natality, and the different story this allows us to tell about selective reproduction.

Birth on the Brink

The emergence of the literature on maternal gifting has coincided historically with the development of reproductive technologies that enable prospective parents to discriminate among embryos in order to give birth to children of their choosing in certain respects. To be sure, when parents are able to avoid giving birth to infants who would be medically condemned to the briefest and most painful lives, such discrimination must surely be recognised as humane and defensible. Indeed, we might say it would be an expression of profound ethical responsibility to the maternal encounter with the natal other. What assisted reproductive technologies do is to stage and perform that encounter in new and different ways—proffering novel and more predictable outcomes—and there is much illuminating empirical research on how this is taking effect on the subjects involved.[12] We might reflect that the medicalisation of pregnancy and childbearing, which couches the maternal encounter with the other in the language of therapy and medically prescribed determinations of beneficence, has tended to transform and subject that encounter to a paternalistic medical gaze. Yet it is the nature of the changes in relation to giftedness that I want to interrogate, since, as we have seen, there is a persuasive thought that being gifted informs the ethical subjectivity that moves us to respond to the demand of the other.

The ethical call to be like a maternal body—like the maternal body that gifts us—runs immediately into the modern subjection of mothers to the clinical practices and moralistic choice regimes that dictate the beneficence or blameworthiness of their maternity. This certainly translates into a regrettable stigma for some mothers who want to treat their maternity more as an ethical encounter than a medical one, those who perhaps want to preserve an unconditional welcome and deliver an uncharted gift of life to another; those who feel that something may be missed or lost in adhering to a liberal eugenic choice regime—or indeed the exercise of any choice at all—in giving life to their child. Recovering the history and thought of the maternal gift of the other, following Guenther's reading of Levinas, may help us to see better or make sense of what might be lost when procreation and motherhood is ordered by a principle of beneficent selection. In terms of the gifting that goes on in maternity, the difference that reproductive technologies makes is very clear: we have gone from gifting in ignorance of the recipient, to gifting more purposively to an other in whom some trait or quality has been secured—that is, from an absolute stranger to a chosen familiar. We are passing from being insensible to the makeup of future generations to being apt to negotiate their characteristics, and while this is morally a triumph for some and yet a disaster for others, I want to understand whether this does significant harm to the relation between past and future.

I have argued that the contemporary bioethical appeal to procreative beneficence in selective reproduction rests on the narrative of a novel and medically-enabled parental capacity to overcome the anarchic and profligate course of natural reproduction and the maternal body—to act newly as gift-givers rather than be in passive receipt of whatever lives arise out of an amoral creation lottery. This narrative colludes uncritically with the historical intellectual erasure of, or failure to see, that we are all given life subject to, from and with mothers, and out of a certain essential and originary generosity of air, time and flesh of maternal bodies that fundamentally resists reciprocation in kind. There is, then, a counter narrative to be told about selective reproduction that acknowledges the immemorial gifting of the maternal body, which tells of a shift not from passive receipt to beneficent giving, but from one kind of gift, power and beneficence to perhaps another. I have attempted in this chapter to draw out what kind of gift the maternal gift of life is, or has been; what it is that the gift gives, and why it may be significant for understanding subjectivity and the way we relate to one another. Some recent feminist philosophy of the maternal suggests compellingly that maternal thinking, drawn from maternal facts, opens up new ways of thinking about ethics and exchange that present more hospitable and perhaps redemptive ways of acting socially and politically. In the final chapter I will consider whether the technological alteration of certain maternal facts might impair or affect the ethics that emerge from it, and the society we create as a result; what changes when being 'like a maternal body' means encountering the predicament of selective reproduction rather than gifting freely and unconditionally to the next generation.

Notes

1. See, for instance, Weiner, A. B. (1976) *Women of value, men of renown: new perspectives in Trobriand exchange.* Austin; London: University of Texas Press. Also O'Grady, K. (2013) Melancholia, Forgiveness and the

Logic of the Gift. *Women and the Gift.* M. Joy. Bloomington: Indiana University Press: 101–110 (Weiner 1976; O'Grady 2013).

2. Perhaps most famous for his emphasis on death would be Martin Heidegger, who espoused Being-toward-death as the most authentic mode of existence; see Heidegger, M., J. Macquarrie and E. S. Robinson (1962) *Being and Time* [Translated by John Macquarrie & Edward Robinson]. (First English edition.). London: SCM Press (Heidegger et al. 1962).

3. John's Gospel, for instance, tells that "Greater love hath no man than this, that a man lay down his life for his friends" (John 15:13). For more on the 'deathly imaginary' of Western thought, see Grace Jantzen (1998) *Becoming Divine: Towards a Feminist Philosophy of Religion,* Manchester: Manchester U.P. (Jantzen 1998).

4. See Derrida (1999) "On the Gift: a discussion between Jacques Derrida and Jean-Luc Marion, moderated by Richard Kearney." *God, the Gift and Postmodernism.* J. D. Caputo and M. J. Scanlon. Bloomington: Indiana University Press (Derrida 1999).

5. In a more persuasive and less essentialist narrative of sexual difference as the source or agent of the gift of reproduction, see Muers's thesis that it takes "at least two" to reproduce: Higton, M. and R. Muers (2012) *The Text in Play: Experiments in Reading Scripture.* Eugene: Cascade Books (Higton and Muers 2012).

6. In addition to Guenther on reading Heidegger after Cavarero, see for additional critical analysis Alison Stone (2010) "Natality and mortality: rethinking death with Cavarero." *Continental Philosophy Review* 43(3): 353-372 (Stone 2010).

7. Imogen Tyler has recently cautioned against the repetition of the maternal as abject in feminist theoretical writing; see Tyler, I. (2009) "Against abjection." *Feminist Theory* 10(1): 77–98 (Tyler 2009).

8. See, for instance, Christine Battersby (1998) *The Phenomenal Woman: Feminist Metaphysics and the Patterns of Identity.* Oxford: Polity. Iris Marion Young (1990) *Throwing Like a Girl and other essays in Feminist Philosophy and Social Theory,* Bloomington: Indiana University Press (Battersby 1998; Young 1990).

9. Frans de Waal is carrying out important and fascinating work in the field of primatology looking at the role of maternity and gratitude in the development of pro-social behaviour, e.g. De Waal (2013) *The Bonobo and the Atheist: In Search of Humanism among the Primates,* New York & London: W.W. Norton & Company.

10. See, for example, Stella Sandford (2001) *Masculine mothers? Maternity in Levinas and Plato. Feminist interpretations of Emmanuel Levinas.* T. Chanter. Pennsylvania: Pennsylvania State University Press (Sandford 2001).

11. Interestingly, a recent study indicated that the brains of fathers acting as primary care-givers alter and adapt to perform and react like those of mothers: Abraham, E., T. Hendler, I. Shapira-Lichter, Y.Kanat-Maymon, O. Zagoory-Sharon and R. Feldman (2014) "Father's brain is sensitive to childcare experiences." *Proceedings of the National Academy of Sciences of the United States of America* 111(27): 9792 (Abraham et al. 2014).

12. See, for example, Franklin, S. (1997) *Embodied Progress: A Cultural Account of Assisted Conception.* London; New York: Routledge. Franklin, S. and H. Ragon (1998) *Reproducing Reproduction: Kinship, Power, and Technological Innovation.* Philadelphia: University of Pennsylvania Press. Franklin, S. and C. Roberts (2006) *Born and Made: An Ethnography of Pre-implantation Genetic Diagnosis.* Princeton, N.J; Oxford: Princeton University Press (Franklin 1997; Franklin and Ragon 1998; Franklin and Roberts 2006).

References

Abraham, Eyal, Talma Hendler, Irit Shapira-Lichter, Yaniv Kanat-Maymon, Orna Zagoory-Sharon, and Ruth Feldman. "Father's Brain Is Sensitive to Childcare Experiences." *Proceedings of the National Academy of Sciences of the United States of America* 111, no. 27 (2014): 9792.

Battersby, Christine. *The Phenomenal Woman: Feminist Metaphysics and the Patterns of Identity.* Oxford: Polity, 1998.

Benhabib, Seyla. *Situating the Self: Gender, Community and Postmodernism in Contemporary Ethics.* Cambridge: Polity Press, 1992.

Cimitile, Maria. "The Gift of Being, Gift of World(S): Irigaray on Heidegger." In *Women and the Gift: Beyond the Given and All-Giving,* edited by Morny Joy. Indiana: Indian University Press, 2013.

Cixous, Hélène, and Catherine Clément. *The Newly Born Woman* [in Translation of: La jeune née.]. Theory and History of Literature (Manchester University Press). Manchester: Manchester University Press, 1986.

De Waal, Frans. The Bonobo and the Atheist: In Search of Humanism among the Primates. New York & London: W.W. Norton & Company, 2013.

Derrida, Jacques. *Given Time.* Chicago: University of Chicago Press, 1992.

———. "On the Gift: A Discussion between Jacques Derrida and Jean-Luc Marion, Moderated by Richard Kearney". In *God, the Gift and Postmodernism*, edited by John D Caputo and Michael J Scanlon. Bloomington: Indiana University Press, 1999.

Diprose, Rosalyn. *Corporeal Generosity: On Giving with Nietzsche, Merleau-Ponty, and Levinas*. Albany; [Great Britain]: State University of New York Press, 2002.

Franklin, Sarah. *Embodied Progress: A Cultural Account of Assisted Conception*. London; New York: Routledge, 1997.

Franklin, Sarah, and Helena Ragon. *Reproducing Reproduction: Kinship, Power, and Technological Innovation*. Philadelphia: University of Pennsylvania Press, 1998.

Franklin, Sarah, and Celia Roberts. *Born and Made: An Ethnography of Pre-Implantation Genetic Diagnosis*. Princeton, N.J; Oxford: Princeton University Press, 2006.

Guenther, Lisa. "Being-from-Others: Reading Heidegger after Cavarero." *Hypatia* 23, no. 1 (2008): 99–118.

———. *The Gift of the Other: Levinas and the Politics of Reproduction*. Albany: State University of New York Press, 2006.

Heidegger, Martin, John Macquarrie, and Edward Schouten Robinson. *Being and Time ... Translated by John Macquarrie & Edward Robinson. (First English Edition.)*. London: SCM Press, 1962.

Higton, Mike, and Rachel Muers. *The Text in Play: Experiments in Reading Scripture*. Eugene: Cascade Books, 2012.

Hird, Myra. J. "The Corporeal Generosity of Maternity." *Body and Society* 13, no. 1 (2007): 1–20.

Hyman, Gavin. "Disinterestedness: The Idol of Modernity." Chap. 2 In *New Directions in Philosophical Theology: Essays in Honour of Don Cupitt*, edited by Gavin Hyman, 35–52. Aldershot: Ashgate, 2004.

Irigaray, Luce. *The Forgetting of Air in Martin Heidegger*. Constructs Series. 1st ed. Austin: University of Texas Press, 1999.

Irigaray, Luce, and Friedrich Wilhelm Nietzsche. *Marine Lover of Friedrich Nietzsche; Translated by Gillian C. Gill*. New York: Columbia U.P., 1991.

Jantzen, Grace. *Becoming Divine: Towards a Feminist Philosophy of Religion*. Manchester Studies in Religion, Culture and Gender. Manchester U.P., 1998.

Kant, Immanuel, and Mary J. Gregor. *Critique of Practical Reason* [in Translation of Kritik der praktischen Vernunft.]. Cambridge: Cambridge University Press, 1997.

Kristeva, Julia. *Powers of Horror: An Essay on Abjection*. New York: Columbia University Press, 1982.

Levinas, Emmanuel. *Otherwise than Being, or, Beyond Essence* [In Translation of Autrement qu'etre.] Pittsburgh: Duquesne University Press, 1981.

Levinas, Emmanuel, and Alphonso Lingis. *Totality and Infinity. An Essay on Exteriority ... Translated by Alphonso Lingis.* Pittsburgh: Duquesne University Press, 1969.

Mauss, Marcel. *The Gift: The Form and Reason for Exchange in Archaic Societies* [in Translated from the French.]. London: Routledge Classics, 2002.

Milbank, John. *Being Reconciled: Ontology and Pardon.* Radical Orthodoxy Series. London: Routledge, 2003.

———. "The Ethics of Self-Sacrifice." *First Things*, no. 91 (1999): 33–38.

Muers, Rachel. *Living for the Future: Theological Ethics for Coming Generations.* London; New York: T & T Clark, 2008.

O'Grady, Kathleen. "Melancholia, Forgiveness and the Logic of the Gift." In *Women and the Gift*, edited by Morny Joy, 101–110. Bloomington: Indiana University Press, 2013.

Oliver, Kelly. "The Maternal Operation." In *Derrida and Feminism*, edited by Feder Rawlinson & Zakin, 53–69. London and New York: Routledge, 1997.

Parker, Andrew. *The Theorist's Mother.* Durham: Duke University Press, 2012.

Protevi, John. *Political Physics: Deleuze, Derrida and the Body Politic.* Transversals: New Directions in Philosophy. London: Athlone Press, 2001.

Ruddick, Sara. *Maternal Thinking: Towards a Politics of Peace.* London: Women's Press, 1989.

Sandford, Stella. "Masculine Mothers? Maternity in Levinas and Plato." In *Feminist Interpretations of Emmanuel Levinas*, edited by Tina Chanter. Pennsylvania: Pennsylvania State University Press, 2001.

Simmel, Georg, and Kurt H. Wolff. *The Sociology of Georg Simmel.* Glencoe, Ill.,: Free Press, 1950.

Stone, Alison. "Natality and Mortality: Rethinking Death with Cavarero." *Continental Philosophy Review* 43, no. 3 (2010): 353–372.

Tyler, Imogen. "Against Abjection." *Feminist Theory* 10, no. 1 (2009): 77–98.

Weiner, Annette B. *Women of Value, Men of Renown: New Perspectives in Trobriand Exchange.* Austin; London: University of Texas Press, 1976.

Young, Iris Marion. *Throwing Like a Girl and Other Essays in Feminist Philosophy and Social Theory by Iris Marion Young.* Bloomington, Ind.: Indiana University Press, 1990.

Walker, Michelle Boulous. Philosophy and the Maternal Body: Reading Silence. London; New York: Routledge, 1998.

5

Natality and Generations

Abstract This final chapter reconsiders selective reproduction as occasioning a shift from one kind of maternal giving to another, rather than a novel form or capacity for giving enabled by reproductive technologies. In this light, feminist correctives to Arendt's philosophy of birth, or natality, and its bearing on education, caution against reproductive practices that seek to determine the lives of future generations, but rather preserves their freedom to renew the world in unpredictable ways. A more hospitable notion of generational beneficence is therefore suggested, which holds open the experience of reproduction as the site of expressing an ethical response to natality, where refusal to choose one's children can be one virtuous response.

Keywords Gift · Natality · Freedom · Education · Generations · Birth

I have argued that the normative choice regime being constructed around selective reproduction tacitly mobilises a discourse of the gift to commend an ethic of Procreative Beneficence and the higher virtue of parents who select embryos and children with the greatest foreseeable advantages in life. My contention is that this narrative of a newly

© The Author(s) 2017
S. Reader, *The Ethics of Choosing Children*, Palgrave Studies in Ethics
and Public Policy, DOI 10.1007/978-3-319-59864-2_5

acquired capacity for beneficence in reproduction trades on a rejection or denial of the basic, embodied generosity of the maternal body; beneficence and gift are written into the discourse of procreation only with the capacity for purposive selection and against the untrustworthy and perilous body of the mother. While the liberalism reflected in contemporary bioethics posits a progressive moral departure from the past, then, it still seems to inherit a derogation of the 'natural' and maternal source of life that is as old as Plato. I have suggested that at least one reason for that persistent negation of the maternal, or even matricide, is that *being* the gift of another, fundamentally dispossessed of the possibility of reciprocation, is conceived as an impossible arrears that cannot be repaid. More recent feminist philosophies of the maternal have sought to draw out and affirm the importance of that state, our corporeal and relational dependence upon others, as vital to ethical self-understanding, perhaps 'moral memory' or even an innate pro-sociality. Yet they also offer us a revision of the bioethical narrative that renders selective reproduction as auguring a shift from receiving to gifting the lives of our children; that is, it marks, rather, a shift from giving—air, time, corporeality—in unknowing hospitality to strangers, to giving conditionally to those selected in virtue of some quality they can be predicted to possess or not to possess.

In this final chapter of the book I will consider how this revision to the orthodox bioethical account of the maternal body allows for a different ethical perspective on selective reproduction. For this we engage with the work of Hannah Arendt, and her idea of natality, conceiving birth as the central category of political thought. Although Arendt famously reconfigured political thought to place birth at its centre, she paid remarkably little attention to the maternal phenomenology of birth and the responsibilities it engages with respect to future generations. Rather, she focused on education, or nurture, as the fundamental site of the expression of our attitude to the fact that the world is renewed by the new and unknowable beginning that arrives with each child. As such, she overlooked the connection between the event of that embodied gift of birth and the principled freedom she demanded for future generations to renew the world in ways unforeseen by us. The same responsibilities that she prescribed in the moral practice of education then, in passing the world of human affairs from generation to generation, I will suggest extend to the consideration of beneficence in the practice of choosing children. Returning to the arguments of reproductive bioethics, I will suggest the idea of a generational beneficence, which pursues and expresses the same responsibilities that are at stake in the educational task of passing a world more or less well between past and future.

The beneficence of the reproductive selections that are commended to prospective parents are, as we considered in Chap. 1, generally evaluated insofar as they create an 'open future' for the child. The ideal is that individuals ought to be created capable to direct their own lives in whichever way they choose, to be unlimited in their life choices by inherited privations or disadvantages that frustrate or prohibit certain possibilities. Yet all lives remain fundamentally limited to being in the world with the particular others and at the particular time we exist; we are bound to share with the generation around us the circumstances of the world we inherit and thereafter shape for ourselves. As Heidegger wrote, "Our fates have already been guided in advance, in our Being with one another in the same world and in our resoluteness for definite possibilities…Dasein's fateful destiny in and with its 'generation' goes to make up the full authentic historicising of Dasein" (1962, 436). By emphasising the fact that we are historical beings, located at a specific time in common with our generation, and therefore limited to a specific range of potential experiences, Heidegger began to address the contingency and finitude of the human condition that had been so long resisted by Platonic and major religious traditions of thought. It is here that we see the roots of a new way of thinking about the human condition through an appreciation of birth, although for Heidegger it was the deathly limit of our finitude that remained the focus of his thinking.

In language that prefigures the creation lottery trope of contemporary bioethicists, Heidegger refers to our condition of 'thrownness' into the world and towards death. It is the premonition of one's "ownmost" death that, for Heidegger, equips us with the possibility of authentic existence, of prising our individual lives away from others who cannot die my death for me. In this way, we can think of it as an affirmative and almost revelatory "freedom towards death" (311) that releases me from the yoke of society and stimulates me to act against the norms and fashions of others according to my own will. But although he focuses upon the future-oriented destination of Being, Heidegger's analysis implies with it a consideration of natality in the condition of being 'thrown' into a world to begin with. While he draws out the loneliness of Being-towards-death, following philosopher Dilthey and Rickman (1976), Heidegger is also concerned to identify us in our historicity, the fact that we are "the sort of beings who are born into a world that is already old…Our thrownness into the world is what makes us historical beings, and the response that it requires of us is a decision—or the avoiding of a decision—about how to receive what is handed down by those who went before" (O'Byrne 2010, 5–6). So while he prefers to elevate the individuating death to which Being projects us, what is also fundamentally implied in the thrownness of Being is receiving life from and with others—a decision about how to receive the world from the past, then also how to hand it down to the future.

With his language of thrownness, Heidegger typically avoids the basic physical fact, language and significance of being born of a woman (see also Irigaray 1999). At a remove from metaphysics, however, sociologist Karl Mannheim brings us closer by recognising that "the sociological phenomenon of generations is ultimately based on the biological rhythm of birth and death" (1952, 290). That is to say, new participants in the cultural-historical process are constantly emerging through birth, while former participants in that process are continually disappearing by death. This creates the predicament of belonging to a generation, and a temporally limited part of the historical process. For Heidegger, this meant a focus upon death as delivering meaning, but Mannheim shifts us towards acknowledging the political and ethical significance of birth as well. For if there were no change of generation, and the world and human culture were only inhabited and developed by the same individuals, any established social pattern would likely just be perpetuated, along with the mistaken or even harmful ideas and practices built into it. It is only with the emergence of brand new individuals in birth, who assimilate and interpret the world anew, that social patterns and attitudes are challenged and changed. As Mannheim wrote, "with the advent of the new participant in the process of culture, the change of attitude takes place in a different individual whose attitude towards the heritage handed down by his predecessors is a novel one" (294). It is only the

change of generation, and what Mannheim terms the radical 'fresh contact' provided by novel newcomers to existence, which promises cultural
and political change, teaching us "to forget that which is no longer useful
and to covet that which has yet to be won" (294). Only the continuous
emergence through birth of new actors in the cultural process secures the
prospect of revision to that process, of social change and political reform
of the way in which we inhabit this world together. It is in relation to
the sociological phenomenon and political negotiation of generations
that I wish to consider anew the moral relevance of choosing children,
this rhythm of the given world and the importance of 'fresh contact' suggested by attention to being given by maternal bodies.

Education and the Problem of Generations

Mannheim spelled out the tension that follows from the continual
emergence of new actors who enter the world by birth, meeting in
their radical novelty with the cultural heritage passed down by the pre-
existing inhabitants of the world they are born into. Although various
economic and social factors now allow us to draw certain distinctions
between generations,[1] the transition from one to the next is continuous, so that generations are actually "in a state of constant interaction"
(1952, 301). It is necessary to perpetually transmit an accumulated cultural heritage to the next generation who arrive as a constant stream
of strangers into our world, yet as these generations abide in the same
space for a time the tensions between them can be acute. According to
Mannheim, while younger generations are closer to 'present' problems
and more dramatically aware and invested in a process of destabilisation, older generations "cling to the re-orientation that had been the
drama of *their* youth" (301). We needn't look too hard to see signs of
this strain, whether gradually through shifts in social attitudes and life-
styles as we age, or more dramatically in revolutionary political move-
ments or resistances. These encounters, manifold and everyday, intimate
and public, all entertain that friction or antagonism that arises from the
'fresh contact' of a new generation that destabilises and challenges an
established or inherited order. For Mannheim "This tension appears

incapable of solution except for one compensating factor: not only does the teacher educate his pupil, but the pupil educates his teacher too" (301). That is, there is a balance to be found in the encounter between old and new, between the representatives of an accumulated culture and the 'fresh contact' of youth, which demands of both parties a willingness or openness to be educated by the other. It is an ethical demand that requires intelligence and vital sensitivity to any relation and advantage of power that exists between the parties, particularly when extreme youth, and indeed old age, are factors in the encounter.

When Hannah Arendt famously diagnosed a loss of authority in the modern age, she identified education as a calamitous casualty of that loss, echoing Mannheim's observations more critically by stressing the need for "the continuity of an established civilisation which can be assured only if those who are newcomers by birth are guided through a pre-established world into which they are born as strangers" (1961, 92). Essential to this task, she argued, is an assumption of responsibility for the world beyond a qualification for teaching, but as representatives of the world to newcomers even though one may wish it were other than it is. Education is the point at which state institutions can exercise purposive and directive influence upon the next generation, where parenting has traditionally also taken or collaborated in this responsibility, both to preserve and hand down our inherited knowledge about what the world is like *and* to leave the future open to renewal by the next generation. That is, in Arendt's words, "where we decide whether we love our children enough not to expel them from our world and leave them to their own devices, nor to strike from their hands their chance of undertaking something new, something unforeseen by us" (196).

It is the Jesuits who are, of course, famed for the motto 'Give me a child until he is seven and I will give you the man,' which demonstrates neatly the influence that one generation can exert over the next. The asymmetry of the adult-child relationship can stymie a child's openness to alternative ideas, philosophies and opportunities, and their own intellectual freedom and imagination to revise and re-envision the world for themselves. Arendt points out that one can hardly avoid the danger: "It is in the very nature of the human condition that each new generation grows into an old world, so that to prepare a new generation for a new world can only mean that one wishes to strike from the newcomers' hands their own chance at the new" (177). Arendt helps us to recognise that in negotiating and overseeing the introduction of newcomers to the world, there is a tension between preserving what we value of it against the destabilisation of the new and leaving it open to the citizens of the future. Just as Mannheim suggests we can be open to learning from and being changed by the agency of the 'fresh contact' that arrives with each new generation, so Arendt suggests that education should balance a responsibility for instructing children about the world against preserving the capacity of newcomers to usher in new and unforeseen ideas of what the world might be like otherwise. Politically, and pedagogically, there is an imperative to safeguard liberty for newcomers to set the world right where the values and visions of the old may be outmoded, pernicious or too modest. But as the motto of the Jesuits makes clear, the tempting and estimable return for the labour of rearing a child is the character and perhaps political/religious allegiances of the adult. The following statement by Arendt is crucial:

> The problem is simply to educate in such a way that a setting-right remains actually possible, even though it can, of course, never be assured. Our hope always hangs on the new which every generation brings; but precisely because we can base our hope only on this, we destroy everything if we so try to control the new that we, the old, can dictate how it will look. (192)

Save for more violent means, the sole manner in which human beings have been able to control the new, until now, is by transmitting selectively and conditionally through education and nurture *only* that which they value or honour about their culture, to the exclusion of alternative ideas and indeed a receptiveness to doubt and conjecture. Setting out in education purely to discipline what Heidegger described as the resoluteness of the next generation for the definite possibilities that *we* have prescribed in advance defeats a hope for a renewal and continual 'setting right' of the world we are handing down.

Mannheim and Arendt point us towards an appreciation of the demand to educate in such a way as to leave the child's future open to new attitudes, values and politics in adulthood—not just for their own individual freedom and benefit, but for a collective renewal of human culture and civilisation. For Arendt the stakes could hardly be higher, then. The asymmetry of the relation in which that demand is entertained—between adult and child, host and stranger—creates the grave danger and possibility of using or withholding education, however, to regulate unpredictability and shut down the capacity of future generations to refashion the world and initiate something new. One key sign of the abdication of responsibility in education, for Arendt, is teaching children less what the world is like, and more what she calls "the art of living" (195). That is, how to simply make an individual success of one's own life within the definite possibilities of an anticipated order, as opposed to understanding and setting right the world to expand the possibilities for living together well in plurality. In education we should both inherit and transmit responsibility for the renewal of the world, as well as for the livelihood of the child. What concerns Arendt as, she argues, should concern us all "is the relation between grown-ups and children in general or, putting it in even more general and exact terms, our attitude toward the fact of natality: the fact that we have all come into the world by being born and that this world is constantly renewed through birth" (196). As I shall explain in the next section, though, for all such talk Arendt was relatively unconcerned with the politics and phenomenology of birth itself. However, we will see how the new practices of selective reproduction become implicated in the very same concerns for our natality that she described as so pressing in education. Indeed, we shall then see how the parity presumption of nature and nurture noted in Chap. 1 may caution against selection. Reproduction may become another site of the same tension and ethical encounter as Arendt diagnoses in education, a nexus of responsibility for both the world and the child where natality is at stake, where a danger is that 'we so try to control the new that we, the old, can dictate how it will look.'

Arendt on Natality and Givenness

Hannah Arendt's concept of natality is like a more critical outworking of what Mannheim describes as 'fresh contact' in his account of generations.[2] As the quote above indicates, Arendt employs the term natality to denote both the new beginning contained in the *event* of birth itself, and the *principle* of the human potentiality for renewal of the world through action by virtue of that birth. As Arendt describes in *The Human Condition*: "the new beginning inherent in birth can make itself felt in the world only because the newcomer possesses the capacity of beginning something anew, that is, of acting" (1958, 9). Human action, for Arendt is the highest realisation of human potential, the 'making itself felt in the world' of the capacity for new beginnings that arrives with the birth of a newcomer. It is the activity that occurs directly between people, disclosing our identity as individuals in a plurality of others, creating and renewing a common world, and setting us apart from the lives of non-human animals as historical beings. The freedom to act then, comes not (only), as Heidegger might say, from an individuating premonition of mortality, but from the capacity to begin that derives from the embodied fact or event of birth in which we arrive into the world as beginnings. "Because he *is* a beginning, man can begin; to be human and to be free are one and the same. God created man in order to introduce into the world the faculty of beginning: freedom" (Arendt 1961, 167). Humanity is constantly renewed by birth, which brings new and unique beginnings to bear on a common world that would otherwise be condemned to expire or repeat itself; "historical processes are created and constantly interrupted by human initiative, by the *initium* man is insofar as he is an acting being" (170). The natal quality of human beings, born such that "nobody is ever the same as anyone else who ever lived, lives, or will live" (1958, 8) means that the improbable can be expected in human affairs; the manner in which that condition is negotiated is the challenge for political and ethical relations between the plurality of natal beings. It is not the certainty of our end, but the miracle of our beginning that is the defining condition of the human freedom to act to establish, alter and renew a common world.[3]

Arendt rightly turns contemplation of the human condition away from coming to possess our "ownmost" possibility in death, to appreciating the "infinite improbability" (1961, 169) of our birth as that which equips human beings to act to perform what is infinitely improbable. As Arendt says, this "is possible only because each man is unique, so that with each birth something uniquely new comes into the world.

With respect to this somebody who is unique it can be truly said that nobody was there before" (1958, 178). However, essential for Arendt in proceeding to act in the world is the capacity to intervene with speech and language into the public sphere; the insertion of oneself into the world which is "like a second birth in which we confirm and take upon ourselves the naked fact of our original physical appearance" (176–177). This separation of linguistic from physical natality, where only by virtue of a secondary birth via speech does the subject insert him or herself into the human world as a political actor, clearly echoes the well-rehearsed philosophical effacement of significance in the given maternal origins of our finitude, and the rebirthing of human actors into transcendent or more essential realms of action or states of being. Arendt opposes the 'givenness' of our biological existence with the more crucial capacity for action as unique individuals in the political sphere amongst a plurality of others. It is this secondary birth into language, laid over and after the physical event of natality, which merits the greatest significance for Arendt and allows the familiar exclusion of the maternal gift of birth.

In *The Origins of Totalitarianism* (1967), Arendt reflects that, historically, the highly developed political life breeds "a deep resentment against the disturbing miracle contained in the fact that each of us is made as he is—single, unique, unchangeable" (301). She describes this "dark background of mere givenness" as a threat to the public sphere, which is based on the law of equality, since the contingency of our differentiation in being born as we are "reminds us of the limitations of human activity—which are identical with the limitations of human equality" (301). In most of her writing, Arendt seems to equate givenness with the brute, bare accident of biological life, the necessity of embodiment most appropriate to subsistence and consideration in the private sphere or household. It is out of this 'dark background' that human beings must liberate themselves to participate in the *bios politikos* as equal citizens. As Peg Birmingham observes, it seems remarkable that in so often dismissing mere life, or *zoe*, to the realm of need and necessity, Arendt herself appeared to fall "blindly into the very subterranean stream of Western history that…she had warned her readers about" (2006, 75). Arendt's account of the event of natality plays down its *arche* or origin in physical birth from the maternal body, repeating the relegation of the gift relationship between mother and child to a forgettable but necessary accident that, in this case, prefigures the subsequent and more significant birth of the political actor, or individual. In doing so, she resists an attention to maternity as being the object or site of action that, insofar as it may express an attitude to natality and renewal of the world—as well as being the event of birth that creates it—engages the same responsibilities as she prescribed in education. For Arendt, natality may be the central category of political thought, but the maternal event of birth remains, as Cavarero puts it, "the given of an ontology that is not able to make itself political" (2005, 196).

It is telling, then, that Arendt writes of mere givenness rather than gift, of that which is "mysteriously" given us by birth (1967, 301), rather than of life that is gifted by maternal bodies. Just as Irigaray uncovered the silenced maternal gift of air in Heidegger's thought, so Cavarero has fleshed out Arendt's philosophy of natality by recalling it to the embodied maternal origin of human beginning in birth. As we saw in Chap. 3, Cavarero takes to task the formative Platonic thought that sought to identify in the pseudo-midwifery of philosophy a re-birthing into the perfection of the soul over the body—and a welcome to death as freedom from the bounds of embodiment. As Cavarero acknowledges, Arendt's work "opened up a direction where the figure of the mother can no longer remain invisible" (1995, 7), yet Arendt herself chose to look away, instead distinguishing the labour of reproduction from the political thought of natality. Indeed, where Arendt did reference reproductive technologies in her prologue to *The Human Condition*, she wrote of it as a rebellion "against human existence as it has been given, a free gift from nowhere" (2–3). However, rather than colluding with the notion of birth as a coming from nothing, nowhere, or 'thrownness' into existence, Cavarero emphatically asserts that "birth is a coming from the mother's womb" (1995, 6) restoring maternity and physical birth as essential to what might be more properly described as the gift event of natality.

Natality cannot be divorced or separated from physical birth, as Arendt supposes, to posit a disembodied form of political natality that locates action only in amongst a plurality of actors capable to make themselves felt in the world through speech and language. A baby develops and emerges as unique amongst others, not least to the mother giving life to it. As a corporeal being, its engagement and communication with others is not deferred until the acquisition of linguistic competencies but is rather immediate and embodied, making itself felt in the world to its mother. Cavarero points out, for example, how the voice itself coincides and emerges as one with the body, "alive and bodily, unique and unrepeatable" (2005, 2). Elsewhere, Peg Birmingham (2006) has argued that there can be no neat distinction between the materiality and sociality of birth. Linguistic natality is not tidily and latterly laid over physical natality; rather, as Birmingham observes, "both births are inseparable and always found together" (25).[4] Arendt's separation of physical birth and a principle of natality allows her to leave the figure of the mother in the shadows, and to exclude from her analysis of political life the significance of maternal agency in giving birth to biological life. In maintaining this distinction, Arendt denies the originary event of natality in the gift of the maternal body, and thereby allows her philosophy to collapse into the same errors of patriarchal thought which, as Guenther explains, "maintains the maternal gift at the root of masculine subjectivity—and, arguably, at the root of philosophical discourse—without recognising this gift as such" (2008, 101).

For Arendt, the asymmetry of the relationship to maternity in our 'givenness' or the event of physical natality would be a barrier to the recognition of uniqueness that comes through the disclosure of the self through speech and action in public life. Given her elevation of the public political space as the site for the incarnation of the human actor, it is perhaps no surprise that Arendt overlooks our maternal origins since, as Cavarero points out: "Birth is inevitably a relationship between unequals…The natal relationship of every human to her/his mother is unbalanced since accident, innocence, and indebtedness are intertwined within it" (1995, 81–82). The event of birth as existential gift undermines the political premise of equality that Arendt holds essential to free human activity in the political sphere. Rather, as Cavarero elaborates, it "upholds a relational ontology consisting of unbalanced and even unilateral exposures" (2011, 196). As such, Arendt leaves such unbalanced and unequal exposures in the private sphere, along with the mothers who create them. The gift event of natality, as the arrival of human beings into the world by virtue of birth, jars with the 'horizontal' pattern of relations in the public realm that sustains the liberty of the autonomous actor, given to act freely by their own devices, unencumbered by dependency, volunteering responsibility for others. However, as we discussed in the previous chapter, an affirmative ethical account of being gifted by a maternal body, "not only confirms the relational frame of Arendt's ontology, but also, by predisposing it towards an altruistic ethics, cautions against dreams of horizontal reciprocity and compels us to understand it in terms of dependency" (199). Cavarero's suggestion seems to be that the event of natality conditions us to commit acts without expecting a return, to exercise care for one another confounding reciprocity, to be moral not because we are *dis*interested but because we appreciate that in our dependence and plurality we are profoundly interested in doing good for others. Perhaps to instil the moral memory and inclination to give gifts.

Arendt herself comes close to recognising this in her recommendation of the most appropriate—or perhaps ethical—response to the condition of the "disturbing miracle" (1967, 301) of our givenness in gratitude. Rather than resenting the unchangeable contingency of our birth, she advises that we respond with "a fundamental gratitude for the few elementary things that indeed are invariably given us, such as life itself, the existence of man and the world" (1951, 438). While she is reluctant to identify the source of that givenness in the agency or body of another, like Simmel Arendt does appear to acknowledge the power and necessity of gratitude in orienting us to recognise a responsibility for others, in the shared condition of being born as unique individuals into the same world together. "In the sphere of politics, gratitude emphasises that we are not alone in the world. We can reconcile ourselves to the variety of mankind, to the differences between human beings...only through insight into the tremendous bliss that man was created with the power of procreation, that not a single man but Men inhabit the earth" (439). This seems to open up a space for recognition of the significance of the event of natality, which she overlooks elsewhere; that is, out of the equal contingency of the 'givenness' of the plurality of human actors, we draw a moral sense of a shared predicament and responsibility for the different actors we encounter in the public sphere.[5] By overlooking the maternal phenomenology of the event of natality, perhaps in unease at the asymmetry of our 'givenness,' Arendt then fails, however, to identify that a motivation for being reconciled to plurality may lie in an originary unilateral exposure to the hospitable maternal body, rather than a disembodied and abstracted 'power of procreation.'

One further connection that Arendt fails to make between the event of natality, properly understood, and an attitude towards the task of renewing a common world, is the construction of a *principled* political freedom for future generations precisely on the basis of our appearance as strangers. Arendt writes of education as the point of a decision as to whether we take responsibility for both the child and the world, one that she makes clear can be inspired only by love (1961, 196). That is, she actually offers no principled obligation to educate and nurture children with a view to either protecting the world or preserving a political freedom for future generations to undertake a new beginning unforeseen by us. As Mannheim explained, each generation as they age tends to seek to preserve and protect the cultural constructions and creations of their own making; we have vested interests in safeguarding the order that we have established against the threat of the new, particularly now that increased life expectancy sustains that interest ever longer. The impetus of love alone looks a little fragile against the prolonged vitality and hegemony of current generations, particularly when much of that power is held in accumulated wealth and property. We work hard and often sacrifice a great deal to raise the next generation into adulthood, and the expectation is often for some kind of return upon that investment by way of a child's achievements and shared values. What's more, the norm is now to accept a price of material and financial debt to the state—or even to one's parents—for the privilege of an introduction to the world. The economic premise of that introduction coercively forecloses the political agency to challenge the debt it imposes and the contractual relations it establishes. Where finance in particular holds generations in thrall to the last, this makes for especially effective and pernicious means for striking from the hands of the young their time and chance of undertaking something new in the world.[6]

In the following section, I shall argue that, following Cavarero's correctives to Arendt's thought, an understanding of natality as gift event enables us to make a principled argument for an attitude to the *principium* of natality that grants freedom to the next generation to renew the world as they see fit. We have seen how Arendt's thinking around natality, the capacity to begin, as a basis for politics, radically works against a history of patriarchal thought so preoccupied with death. As Imogen Tyler optimistically states, "Perhaps in Arendt's natal thinking lie the seeds of an alternative, future-orientated politics which might challenge the predominant neo-liberalism" (Tyler 2009, 1). Arendt herself relegated the event of birth and mysterious 'power of procreation' from the political space, while keeping the *capacity* for new beginning that birth engenders for her account of action in the public sphere. Yet that capacity stems directly from the event of birth, so that, as Diprose and Ziarek explain, "the connection between natality and political agency is established, not exclusively in the public realm of "togetherness" or the political, but also via the "first order" of birth" (2013, 111). Diprose and Ziarek argue forcefully why maternity, as an expression of agency and futurity, is, like education, so often therefore the target of regulation by government or religion. The preservation of the world for the expression of natality can be severely jeopardised not just by indifferent or ultra-conservative practices of education then, as Arendt suggested, but also by disciplining or coercing maternity. While it follows from this, of course, that a woman's decision to give or not to give birth should be made as freely as possible, I will suggest that the strangeness and unpredictability of the natal as the recipient of the gift of life, importantly secures an imperative in principle to hold the future open to the expression of the unpredictable individual and collective political agency of future generations. The decision to give life conditionally, or selectively, thereby entertains the individual and collective political danger of holding the future in thrall to the continued expression of the condition under which the gift was given.

Freedom and Gifts to Strangers

"Folk like him loek for a return from their bairns."
Githa Sowerby, *Rutherford and Son* (1912)

"I wish to make sure it goes to the right sort of person."
Tony Hancock, *The Blood Donor* (1961)

Rutherford and Son was a play written in 1912 by Githa Sowerby which explored, among other things, the obligations between parents and children. In this case, the parent is a bull-headed patriarch and capitalist who demands of his son that he carry on the family firm, secure his legacy and give over the secret of an invention that would benefit the business; according to Rutherford, it is only what is owed him by his son in return for bringing him up. While his children aspire to live out dreams and lives of their own, Rutherford's insistence on the rightness of the debt they owe him for raising them eventually drives them all away. In the end, his desperate daughter-in-law Mary drives a bargain with him to keep and clothe her infant son for 10 years, after which she promises to hand his grandson over to be trained up to take on the business. The play makes us think about the limits and privileges of the freedoms we aspire to, of one generation's demands for a return on the gift of raising the next, against the will and capacity of that new generation to express and direct itself differently. Ultimately, it is about the *principium* natality, the father's attitude to the freedom of the son. To be sure, it is familiar enough to feel a sense of obligation to one's parents, but in Sowerby's play we see how relationships break down when these are insisted upon with the unjust moral force and expectation of a settlement or contract.

Many of us are painfully aware of the ways in which we have not lived up to the hopes or expectations of our parents; very often the course we tread is one at wild and dismaying variance to that which had been plotted for us. *Rutherford and Son*, staged at the dawn of an age of unprecedented social mobility and increasing meritocracy, shows us what happens when that variance is acted out at the expense of what the parent feels is a life's work and industry. To be sure, in different times and cultures most people have been more decisively constrained to the

paths set out for them by the requirement to contribute as early as prac-
ticably possible to the household in which they were given birth and the
brief time of childhood. While the capacity to begin and act anew into
the world may be the essence of the human condition, we mustn't pre-
tend that the great majority of human beings have not been violently
compelled to repeat the past out of necessity and obligation. Although
that puts the predicament of modern childhood in perspective, the
increasing economisation of life processes and hyper-parental investment in one's offspring brings significantly constraining obligations to
bear upon the child and their natality. As Jacques Godbout puts it, "The
gift to the child may be the quintessential form taken by the modern
gift, and the debt incurred the most difficult to assume" (1998). Not
only does the relation to the child stand out as the most celebrated form
of oblation, but for the purposiveness, intensity, duration and expense of
modern parental giving, the child finds him or herself more obliged to a
return by way of attainment or compliance than ever before.

One model of giving designed to resist a return from the recipient to
the giver is the system of blood donation in the UK. Although there may
be other compensations, blood donation brings us close to the ideal of
the gift because, in a way, it is a gift that is never received; donor and
recipient are kept apart from one another so there can be no question
of direct reciprocity or of the giver holding the recipient in his or her
debt. It is an arrangement where the strangeness or radical unfamiliarity of the recipient is deliberately built in, making it impossible to insist
upon a return for the gift that one volunteers. As such, there is, as
Titmuss acknowledged in his masterly study, "in the free gift of blood to
unnamed strangers no contract of custom, no legal bond, no functional
determinism, no situations of discriminatory power, domination, constraint or compulsion" (1972, 239). Although the gift may not always
be compelled by altruism, it cannot be motivated by a desire to exercise
over the recipient some obligation to the giver; while the gift may preserve the life of the recipient, they are not then beholden to the giver
for what they do with their life thereafter since at the moment and in
the process of the giving they remain unknown to each other. The donor
gives to whomsoever awaits in need at the time and place of the distribution of their gift, preserved from direct obligation to the giver but free to

perhaps be moved in gratitude, and a spirit of solidarity, to pass on whatever gifts they are able to in their turn (see also Murray 1987).

In a brilliant episode of the 1960s BBC comedy show Hancock's Half Hour, *The Blood Donor*, the eponymous anti-hero's attitude throughout subverts the ideal of disinterested altruism that is supposed to lie behind the act. He attends the donation centre out of a misguided sense of national service, craves some kind of recognition—"I just think we ought to get a badge, that's all"—baulks at the amount of blood required, is persuaded by the prestige he attaches to his rare blood type, and finally ends up receiving the gift of his own blood back after cutting himself on a bread knife, thus defeating the whole idea of the donation. The gift of his blood does not go forward to anyone else, but rather circles back directly to the giver. While the entire system is modelled to preclude the possibility of a return, of the direct reciprocity or indebtedness of another, the hilarious conceit of the episode is that this is really what Hancock is seeking throughout. He telephones the hospital the day after his donation to enquire after the object of his gift, to "make sure it goes to the right sort of person," having failed to appreciate that he would be unable to stipulate the conditions for its distribution, that he had gifted outside an economy of returns with the recipient, who would remain a stranger to him. It is the fact that we can forego reciprocity that elevates the agent and the act to the possibility of being ethical; Hancock's reluctance to do this is what makes him laughable as we see in his motives he is not really being ethical at all, but rather seeks the very situation of "discriminatory power, domination, constraint or compulsion" over the recipient that Titmuss commends the system for ruling out.

Entering into a mode of gifting like blood donation in which the strangeness of the recipient is built in, and then preserved, means that situations of domination or constraint are impossible; the giver is not motivated by power, nor the recipient by compulsion. Until recently, procreation was a mode of gifting in which the strangeness of the recipient *was* built in, although not preserved, rendering the subsequent task of making the natal familiar, as parents and educators, the most crucial site of our attitude to the fact of natality. To passing on the accumulated knowledge and cultural heritage of our families and the world, yet hold-ing the future open for the fresh contact of new generations and, *contra* Rutherford, novel expressions of a world and lives renewed in unforeseen and unexpected ways. However, the strangeness of the newcomer in the gift event of natality, of birth, establishes the rightness of an attitude to the *principium* natality in parenting or education that preserves and encour-ages the individual and political freedom of future generations to under-take something new; to change the world. Because reproduction did not afford the past to consent to the future, it could not hold that future indebted to its designs, either for individual lives or the continuation of convention. The initiation (*initium*) of the relation between a parent and child was a gift event, or welcome, to a stranger, and undertaking to begin a relation on that basis obliges an openness and assent to the unpredictable gifts and talents that newcomer to life arrives with. The new reproductive technologies of diagnosis and selection, however, mean that such relations need not be initiated on that basis, and a legitimate concern is therefore the extent to which parents feel obliged to the unpredictable and unpre-dicted qualities of a child, the extent to which one generation permits the unchecked natality of the next.

When we locate the root of the child's freedom in the gift event of natality, a coming from a mother, we uncover a principled political source of resistance to the yoke of the past over the future, the drive to impress and affirm the regime of one generation into the next. Part of what guarantees and makes that freedom compelling has been the fact that birth has been given only to uniquely new, unexpected and strange others; although life has always been given to the next generation, there must be a moral and political case for freeing that generation from its obligation in receiving the world to fashion it anew precisely because each natal life was not initially scheduled or selected into existence but gifted to strangers, without determining 'the sort of person' they will be. Reproduction was consent to the unknown. Godbout voices the misgivings of countless philosophers before him in proclaiming that "Birth establishes the state of indebtedness as a defining feature of the human condition" (1998, 40). But following our discussions of natality in this chapter, and maternity in the chapters before, however, we can recognise that an unsolicited gift to a stranger cannot justifiably establish a debt that can be held over and constrain the life of the recipient. It is more accurate to state that birth establishes the state of *giftedness* as a defining feature of the human condition—with it come plurality and dependence of course, but crucially the capacity for new beginnings that existing generations can respond more or less openly to. The gift event of natality equips a new generation to reinterpret and recreate a common world differently to the one they inherit, but it is clear that this can be made more or less difficult and approached more or less ethically according to *how* that world is passed down, the attitude of parents, educators and society to the principle of natality.

The selective reproduction or genetic enhancement of children does not alter the *fact* of their natality, their capacity to begin, which, as Guenther elaborates, "arises simply by virtue of being born as this one, with my own beginning in time and my total self-exposure at this initial moment" (2008, 110; see also, O'Byrne 2010). However, the new prenatal familiarisation with the embryo as a condition for welcoming a child into the world, breaks the previously given or natural connection between the event of natality as gift/birth to a stranger and the morally principled expression of natality in preparing a new generation for renewing the world. In breaking that connection it is hard to think that the political imperative is not weakened somehow by being disconnected from a justification rooted in the biology of generation. We can no longer point to the fundamental strangeness of natals, or newcomers to the world, as the basis for seeking to hold the future open for the new beginnings they promise against the old world we have learned to inhabit. When we can choose children, rather, to inhabit the world in the way *we* do, the way *we* think best, or to relate to *us* in the way that we'd like, we can begin to cement that vision into the biology of new generations, constraining the unpredictable differences that move new generations to renew the world in unforeseen ways. We begin to choose children fit for the world we have prepared for them, rather than take responsibility for preparing a world fit for any children.

The same responsibilities that we admire and commend in a rich ethical tradition of hospitality are prefigured in the maternal gift of life, so that we see why Levinas likened the ethical to being like a maternal body.[7] Before the reproductive regimes and technologies that facilitated selection, giving birth was the ultimate gesture of hospitality, the quintessential welcome to strangers upon whom we must rely for the renewal of the world. The admission or invitation of a stranger into one's home immediately implicates a host in responsibility both for the preservation and protection of that home against the potential threat or destabilisation brought with the newcomer, *and* for the newcomer whose uniqueness and unpredictability one assents to in making the invitation. These are the responsibilities that Arendt observes are engaged or must be assumed in the activity of education, of making newcomers to life itself familiar and welcome in the world we have invited them into. Crucial to the engagement of such responsibilities is the subject of the invitation, as an unknown and potentially disruptive stranger, so that we must prepare and offer up our home as fit for whomever they may be; if we can choose who to invite, we need only prepare our home for the particular requirements and preferences they can discernibly be predicted to have. Likewise, when procreation shifts from being gifted to strangers to being chosen for hospitality, when being 'like a maternal body' no longer means assenting to the unpredictable and unknown but rather choosing the subjects of the future to be accommodated within the means of the present, we need not concern ourselves with the collective political task of preparing a world fit for strangers. We need not assume responsibility for the world, or our home, as a place of welcome and hospitality for anyone.

Where procreation was once a gift of life to a stranger, techniques like PGD line up potential natals whose prospective qualities are read off in order to judge which will most likely satisfy the preferences and desires of their parents, and indeed the preferences the future child is likely to possess in our world as we know it. To be sure, such judgements may undoubtedly be made beneficently, that is, according to a parental preference for the child to fare well in the world they inherit. But this brokers a relation between the world and the child where the world is the buyer; the newcomer is selected insofar as they are fit for the common world they will inherit, rather than the world be made fit for whoever may be born into it. As such, the politics and economy of the world into which the natal is born is prioritised over the natality of the child; the tractable newcomer can be bent to the shape and order of the given world, so there is less cause for that world to bend to the shape of the newcomer. The decisions we take and enact ourselves form part of the world that future generations inherit insofar as they express a vision of the good, but the selection of our descendants additionally allows us to etch that vision and expectation into the givenness of those who inherit it. In doing so, we bequeath a performative set of values about what a worthwhile life looks like in our world, and, crucially, close off revisions to those values as we close off the need to challenge them by and on behalf of newcomers whose lives look different. The strangeness of the newcomer, until very recently, has preserved, in principle at least, the liberty of the natal to make of their talents what they will, unindebted to any conscious human agency or design, given life as gifts by maternal bodies. The debate over PGD and selective reproduction seems poorer for failing to recognise that the introduction of that agency does away with such liberty as a principle founded on the erstwhile facts of life.

Conclusion: Towards a Generational Beneficence

Returning to the arguments of bioethics, we can actually affirm, in a sense, what we called the parity presumption in Chap. 1; that is, the moral equivalence of selecting environments or nurture conducing to certain traits, and the possibility of prenatally selecting children for the expression of those traits. Of course, this presumption is usually levelled against objections to selection, but with an Arendtian appreciation of the responsibilities incumbent on parents and educators in their attitude to natality, we reflect quite differently on the obligations that selective reproduction engage. There is parity, or moral equivalence, in the estimation of the attitude to natality that is expressed in the performance of education and selective reproduction. Both child-rearing and now selective child-bearing engage the same question of responsibility and attitude to natality, to the child *and* the world, since they both now *purposively* influence the thought, inheritance and character of the future person. Indeed, Arendt herself fleetingly made the connection between the event and principle of natality when she stridently declared: "Anyone who refuses to assume joint responsibility for the world should not have children and must not be allowed to take part in educating them" (1961, 189). For Arendt, we meet that responsibility in education and child-rearing by introducing newcomers with authority to the world they inherit *and* passing it on to them to express their natality in renewing it for themselves; she is clear that, as the site of negotiating the transition between past and future, education engages obligations both to the child and to the world into which they are born. It must gather up the past and transmit the history and cultural heritage of the world we have in common, but not fetter the future by thwarting the new beginnings that arrive with each generation. It must not programme newcomers to simply repeat the past and just ready them for 'the art of living' in a world they are not responsible for, or hold them hostage to a riveted order by levying debts upon newcomers for the expense or condition of their introduction to the world.

The parity argument, under the permissive presumption of procreative liberty rights, defends those reproductive selections that tend to the same outcomes for the child that we are permitted to seek in moulding the child after their birth, through nurture and education, etc. A revised Arendtian approach to this argument might run rather differently, or at least encompass the additional consideration of insisting on parity with the outcomes *for the world* that we also ought to seek in education. What are our responsibilities to the world in education, and how are they properly honoured or reflected in the responsibilities we assume in our reproductive choices? Our responsibility to the world runs to the collective understandings and traditions that we bequeath to future generations, so that just as we celebrate the plurality of our children's gifts and abilities in education, so we might recognise these in considering procreation. Just as we would not teach that a life with, say, Downs syndrome is regrettable or not worth living, so we should resist the clinical performance of pregnancy with this presumption built in. Rather than be made to feel blameworthy for resisting prenatal testing for such a syndrome, women ought to be able to make decisions about the lives they will gift *with their bodies* according to the kind of world they want to pass on. Just as parents are generally supposed to take a view, concerning a child's education, of the kind of world they want to bequeath, so it should come as no surprise, and without objection, when prospective parents are open to the event of natality as gift to a stranger, assuming responsibility for a world and community hospitable enough for any child who may enter it.

Perhaps it is no coincidence that in an age and culture in which reproduction is subject to greater control than ever before, when the maternal gift of new life can be easily averted, rescheduled and now finally selected, the crisis in education that Arendt diagnosed has developed and intensified. Perhaps there is some link between the increased parental facility to appoint the previously unpredictable 'fresh contact' of new life, and the loss of authority in education and parenting. The more we can take responsibility for the time and life of the child, the less we need to take responsibility for the world itself. Likewise, beneficence in education has become increasingly limited to the preparation of children for gainful employment, qualities of resilience and the acquisition of means to secure a comfortable quality of life. As Mannheim predicted, "The wrong type of democratic education will tend to transform everything into terms of vocational training and adjustment to an industrial order. It will be so concerned to bring about a compatibility with the contemporary that its sense of heritage, or history and tradition, will be cut at the root" (1962, 23). Higher education has also been transformed into a private good now billable to the individual since we suppose they alone will benefit in return—as an individual—from receiving it; the assumption, and indeed possibility, of even desiring to act for love of the world is being stripped out of the system. With this culture of raising young people as the ethical departure point for an argument from moral parity for selective reproduction, parental responsibility is similarly both intensified and limited to ensuring simply that one chooses a child to go well in the world. Procreative beneficence is therefore contained to the remit of securing the best possible life for the child, rather than passing on the best possible world to the next generation. There is a strange reversal here in that we now look to the culture of education and child-rearing as the moral yardstick for regulating the event of natality, rather than take the event of natality to inform our attitude to education and introducing newcomers to the world. It is the consistent patriarchal disavowal of the significance of birth for natality that has allowed this inversion to take place.

Some of the more reductive accounts of human agency to have emerged from the field of evolutionary biology have supported the idea that human beings—or their genes—are motivated fundamentally by the renewal of themselves through sexual reproduction (Dawkins 1976). The growing traction of such a view as a scientific explanation of human behaviour has also perhaps contributed to an indifference to the fact that reproduction also fundamentally renews the world. Not just genes and families, but the world of human society too is renewed in the gift of maternal bodies. Procreation is therefore not just about projecting the best version of ourselves genetically into the future, but also the best version of our world. As well as the harms that might befall any created individual by virtue of our procreative choices, it is legitimate to consider the harms—or benefits—that our decisions write into the collective understanding of the world and the home into which we introduce them. We might therefore suggest that the clinical and cultural performance of pregnancy allows that the scope of beneficence in procreation is understood beyond the given qualities of the child to be inclusive of the kind of world we are giving the child to go well in. While bioethicists like Harris and Savulescu argue that we have compelling moral reasons to condition a welcome to natals according to their possession of certain desired or desirable traits, there ought to be space for acknowledging that there are legitimate moral reasons to preserve an unconditional welcome to natals according to a vision of a desired or desirable world. A world for which we take responsibility as welcome to strangers and to the new beginnings they bring. We might even suggest the name, if it were helpful, of a general principle of *Generational Beneficence*; not opposed necessarily to any other principle but as a broader moral and political responsibility in which selective reproduction, like education, is implicated, or cannot be viewed in isolation from.

To be clear, I think it is absolutely justifiable for parents to use the technologies of selective reproduction to ensure that they have a healthy child; indeed, it can be an expression of profound responsibility for the natal when very seriously harmful and life-limiting conditions are at stake. By the same token, I would not level any grandiose Arendtian responsibility for the world at prospective parents, and mothers particularly, as they negotiate the process of having children. I readily concede the privilege and distance of my position from this predicament. Yet, like many others, I have difficulty subscribing to an ethics or culture that limits an appreciation of beneficence in procreation *only* to the fitness of the child, and is blind to the beneficence in the freedom of the gift itself—or perhaps takes resistance to selective reproduction as indicative of an ethically unsophisticated or uncritically conservative position. In a way, the desire to select our children can actually be seen as expressive of a fundamentally conservative attitude to natality, as it is in education, when we so try to control the new that we, the old, can dictate how it will look. It can be a form of hubris, not in the sense that we assume to appropriate the gifts of nature, but to imply that we esteem our lives so much *just as they are now*, that we do not wish to have them changed or challenged too radically by the arrival of a stranger in the future. As Mannheim remarked of education, "The coat must be cut according to the cloth, but it must also be cut according to the would-be wearer" (1962, 33). When we begin cutting the wearer according to the coat, perhaps it is time to re-evaluate the choices we are making.

When we receive the gift of our lives as strangers, a parental assent to the unpredictable is built into the act; with it, a moral imperative to hospitality for the new nature and qualities of the child, and a principled moral freedom for the next generation to renew the world in ways unforeseen, uncontrolled and undesignated by the last. It is naive to think that the shift enacted by choosing children for the ways in which they are familiar to us and our world, rather than admitting strangers and the responsibility for making them familiar with the world *as they come*—open to their natality—is sociologically and morally innocuous, or else an unqualified good. The nature of what Mannheim called the biological rhythm of generations is subtly changed as generations of newcomers become subject to ever greater control.

Returning to the words of Andrew Solomon, whose work we quoted at the beginning of this book: "Parenthood abruptly catapults us into a permanent relationship with a stranger, and the more alien the stranger, the stronger the whiff of negativity" (2012, 1). In allowing parents to choose children less strange, less mysterious in their conception, selective reproduction alters the task required of one generation in preparing and imparting the world to the next. It allows us to mitigate the 'negativity' that we might encounter and yet still be responsible for, or even learn to love. We must leave open the possibility that our understanding of beneficence may be refined or revised by future generations, and not transmit, either through education or selective reproduction, limits of both the will and the requirement to be more responsible for the world.

Notes

1. E.g. 'Baby boomers' Generation X & Y, Millenials, etc.; see David Willetts' (2010) *The Pinch: How the Baby Boomers took their Children's Future—and why they should give it back*. London: Atlantic Books (Willetts 2010).
2. Although Arendt was familiar with Mannheim, and he is said to have been an influence upon her thought on the social in particular, there is no indication that she directly drew upon his work on generations. For an account of Mannheim's influence, see Walsh, P. (2014) Hannah Arendt on the Social, *Hannah Arendt: Key Concepts*. P. Hayden. Abingdon: Routledge: 124–137 (Walsh 2014).
3. For more on the features of natality, and its repression by a deathly Western symbolic, see Grace Jantzen (2004) *Foundations of Violence*. London: Routledge (Jantzen 2004).
4. Kelly Oliver also advances a similar argument, see Oliver (1997) *Family Values: Subjects between Nature and Culture*, New York; London: Routledge (Oliver 1997).
5. See also Peg Birmingham's work on givenness in Arendt and the development of a basis for human rights: Birmingham (2006) *Hannah Arendt & Human Rights: The Predicament of Common Responsibility*. Bloomington: Indiana University Press (Birmingham 2006).
6. For a particularly compelling account of this, see David Graeber's (2011) *Debt: The First 5000 Years*, New York: Melville House (Graeber 2011).
7. To engage with the most recent and important work of that tradition, see Jacques Derrida's (1999) *Adieu to Emmanuel Levinas*, Stanford: Stanford University Press, and (2000) *Of hospitality: Anne Dufourmantelle invites Jacques Derrida to respond*, Stanford: Stanford University Press (Derrida 1999, 2000).

References

Arendt, Hannah. *Between Past and Future*. London: Faber, 1961.
———. *The Burden of Our Time*. Secker & Warburg, 1951.
———. *The Human Condition*. 2nd ed. [i.e. reissued with improved index and new introduction by Margaret Canovan] ed. Chicago; London: University of Chicago Press, 1998, 1958.
———. *The Origins of Totalitarianism*. 3rd ed. London: Allen & Unwin, 1967.
Birmingham, Peg. *Hannah Arendt & Human Rights: The Predicament of Common Responsibility*. Bloomington: Indiana University Press, 2006.
Cavarero, Adriana. *For More Than One Voice: Toward a Philosophy of Vocal Expression*. Stanford, California: Stanford University Press, 2005.
———. "Inclining the Subject." In *Theory after 'Theory'*, edited by Jane Elliott and Derek Attridge, 194–204. London: Routledge, 2011.
Dawkins, Richard. *The Selfish Gene*. Oxford: Oxford University Press, 1976.
Derrida, Jacques. *Of Hospitality: Anne Dufourmantelle Invites Jacques Derrida to Respond*. Cultural Memory in the Present. Stanford U.P., 2000.
———. *Adieu to Emmanuel Levinas*. Stanford, California.: Stanford University Press, 1999.
Dilthey, Wilhelm, and H. P. Rickman. *Selected Writings*. Cambridge; New York: Cambridge University Press, 1976.
Diprose, Rosalyn, and Ewa Płonowska Ziarek. "Time for Beginners: Natality, Biopolitics, and Political Theology." *philoSOPHIA* 3, no. 2 (2013): 107–20.
Galton, Ray, and Alan Simpson. "The Blood Donor." BBC, 1961.
Godbout, Jacques T., and Alain Caille. *The World of the Gift* [in Translation of: L'esprit du don.]. Montreal; London: McGill-Queen's University Press, 1998.
Graeber, David. *Debt: The First 5,000 Years*. New York: Melville House, 2011.
Guenther, Lisa. "Being-from-Others: Reading Heidegger after Cavarero." *Hypatia* 23, no. 1 (2008): 99–118.
Heidegger, Martin, John Macquarrie, and Edward Schouten Robinson. *Being and Time … Translated by John Macquarrie & Edward Robinson. (First English Edition.)*. London: SCM Press, 1962.
Irigaray, Luce. *The Forgetting of Air in Martin Heidegger*. Constructs Series. 1st ed. Austin: University of Texas Press, 1999.
Jantzen, Grace. *Foundations of Violence*. London: Routledge, 2004.

Mannheim, Karl, and Paul Kecskemeti. *Essays on the Sociology of Knowledge …* *Edited [and Translated] by Paul Kecskemeti*. London: Routledge & Kegan Paul, 1952.

Mannheim, Karl, and William Alexander Campbell Stewart. *An Introduction to the Sociology of Education*. London: Routledge & Kegan Paul, 1962.

Murray, Thomas H. "Gifts of the Body and the Needs of Strangers." *The Hastings Center Report* 17, no. 2 (1987): 30–38.

O'Byrne, Anne E. *Natality and Finitude*. Bloomington: Indiana University Press, 2010.

Oliver, Kelly. *Family Values: Subjects between Nature and Culture*. New York; London: Routledge, 1997.

Solomon, Andrew author. *Far from the Tree: A Dozen Kinds of Love*. New York: Simon and Schuster, 2012.

Sowerby, Githa. *Rutherford and Son: A Play in Three Acts*. [S.l.]: Sidgwick and Jackson, 1912.

Titmuss, Richard Morris. *The Gift Relationship: From Human Blood to Social Policy*. New York: Vintage Books, 1972.

Tyler, Imogen. "Introduction: Birth." *Feminist Review* 93, no. 1 (2009): 1–7.

Walsh, Philip. "Hannah Arendt on the Social." In *Hannah Arendt: Key Concepts*, edited by Patrick Hayden, 124–37. Abingdon: Routledge, 2014.

Willetts, David. *The Pinch: How the Baby Boomers Took Their Children's Future—and Why They Should Give It Back*. London: Atlantic Books, 2010.

Index

© The Editor(s) (if applicable) and The Author(s) 2017
S. Reader, *The Ethics of Choosing Children*, Palgrave Studies in Ethics
and Public Policy, DOI 10.1007/978-3-319-59864-2